T0205374

Studies in Distributed Intelligence

Series Editors

Mohamed Elhoseny, Computers & Information Sci, Apt 142, Mansoura University, Metairie, LA, USA

Xiaohui Yuan, Computer Science and Engineering, University of North Texas, Denton, TX, USA

The ubiquitous sensing, multi-modality data, and various computing platforms enable rapid, personalized, dedicated services for our communities. The global perspectives on the distributed intelligence showcase cutting-edge research that embraces a number of fields from system design to deployment and from data acquisition to analysis in cross-disciplinary areas that require discovery of valuable information from dynamic, massive, diversified data. The research to be included in this series focuses on the pressing, contemporary aspects of the decentralized, computational intelligence techniques for geotechnology applications including environment, resource management, transportation, health, robotics and autonomous vehicles, and security, to name a few.

The series "Studies in Distributed Intelligence" publishes new developments and advances in the areas of distributed intelligence and covers the theories, methods, and applications of distributed intelligence, as embedded in the fields of engineering and computer science as well as the cross-disciplinary fields. The series contains monographs, lecture notes, and edited volumes in distributed intelligence spanning the areas of environmental sensors, big geoscience data analysis, smart sensing networks, distributed systems, edge computing, supercomputing for climate problems, artificial intelligence in environmental monitoring, urban intelligence, self-organizing systems, intelligent transportation systems, soft computing, smart vehicular communication, and hybrid intelligent systems. Of particular value to both the contributors and the readership are the short publication timeframe and the worldwide distribution, which enable both wide and rapid dissemination of the research outputs.

If you are interested in contributing to the series, please contact the Publisher: Aaron Schiller [Aaron.Schiller@springer.com].

More information about this series at http://www.springer.com/series/16393

Xiaohui Yuan • Mohamed Elhoseny

Editors

Urban Intelligence and Applications

Proceedings of ICUIA 2019

 Springer

Editors
Xiaohui Yuan
Department of Computer Science
and Engineering
University of North Texas
Denton, TX, USA

Mohamed Elhoseny
Faculty of Computers and Information
Mansoura University
Mansoura, Egypt

ISSN 2662-3706   ISSN 2662-3714   (electronic)
Studies in Distributed Intelligence
ISBN 978-3-030-45101-1   ISBN 978-3-030-45099-1   (eBook)
https://doi.org/10.1007/978-3-030-45099-1

This Springer imprint is published by the registered company Springer Nature Switzerland AG.
The registered company address is: Gewerbestrasse 11, 6330 Cham, Switzerland

Preface

The proceedings include papers presented at the First International Conference on Urban Intelligence and Applications (ICUIA) held in Wuhan, China on May 10–12, 2019. This conference series provided an international forum to present, discuss, and exchange innovative ideas and recent developments in the fields of computer science, computational geography, and management. The proceedings provide new advancements in theories and inspiring applications to scholars, industry leaders, policymakers, and administrators on the current issues and solutions to support the integration of artificial intelligence into modern urban life and to advance the design and implementation of intelligent utilization and management of city assets.

The ICUIA'19 proceedings include four themes: Technology and Infrastructure for Urban Intelligence, Community and Well-Being of Smart Cities, Smart Mobility and Transportation, and Security, Safety, and Emergency Management, which are well balanced in content and created an adequate discussion space for trendy topics. Two distinguished plenary speakers, Dr. Sos Agaian from City University of New York and Dr. Jinshan Tang from Michigan Technological University, delivered outstanding research outlook and progress in the fields of urban intelligence including "Bioinspired Computational Urban Intelligence: A Long Road Ahead" (Dr. Agaian) and "Object Recognition Based on 3D Shapes Extracted from 3D Imaging and Applications" (Dr. Tang). There were 30 presentations that brought a great opportunity to share their recent research findings. Four industrial partners demonstrated their products of both hardware and software systems during the exhibition as part of the grant event.

The efforts taken by peer reviewers contributed to improving the quality of papers, which provided constructive critical comments. Comments to the submitted papers are greatly appreciated. We are very grateful to the organizing committee members, technical committee members, session chairs, student volunteers, and colleagues who selflessly contributed to the success of this conference. Also, we thank all the authors who submitted paper, because of which the conference became a success. It was their dedication to science and technology and passion to openly communicate with attendees that really made this event fruitful and memorable.

Denton, TX, USA   Xiaohui Yuan
Mansoura, Egypt   Mohamed Elhoseny

Acknowledgments

In recent years, we have observed exponential growth in technology to revolutionize our urban living space for greater convenience, comfort, sustainability, and security. The development of new theories, techniques, and applications as well as the interdisciplinary nature of urban intelligence calls out for a forum to exchange ideas and foster collaborations. The International Conference on Urban Intelligence and Applications (ICUIA) aims at sharing the latest research advancement in algorithms, applications, future directions on artificial intelligence, and urban solutions and bringing together researchers, developers, practitioners, and government officials for sparks and inspirations. The conference provides a unique platform for researchers and professionals to exchange pressing research challenges, novel ideas, and solutions in the smart community and urban space. The success of this conference and quality publication are due to the great efforts of the Committee Chairs and Members including Surapong Auwatanamongkol (National Institute of Development Administration), Khouloud Boukadi (University of Sfax, Tunisia), Mauricio Breternitz (AMD Research), Chris Bryan (Arizona State University), Nelio Cacho (UFRN), Xin Cao (The University of New South Wales), Siobhan Clarke (Trinity College Dublin), Baofu Fang (Hefei University of Technology), Fang Fang (China University of Geosciences), Lichuan Gu (Anhui Agricultural University), Wu He (Old Dominion University), Huawei Huang (Kyoto University), Jinoh Kim (Texas A&M University-Commerce), Yantao Li (Chongqing University), Yu Liang (University of Tennessee at Chattanooga), Jianguo Liu (University of North Texas), Xiao Liu (Deakin University), Yuanyuan Liu (China University of Geosciences, Wuhan), Seng Loke (La Trobe University), Massimo Mecella (Sapienza University of Rome), Nader Mohamed (Middleware Technologies Lab), Sundeep Narravula (Electronic Arts), Barbara Pernici (Politecnico di Milano), Heng Qi (Dalian University of Technology), Juan Antonio Rico Gallego (University of Extremadura), Yonglin Shen (China University of Geosciences, Wuhan), Domenico Talia (University of Calabria), Predrag Tosic (Washington State University), Traian Marius Truta (Northern Kentucky University), Genoveva Vargas Solar (CNRS-LIG-LAFMIA), Iraklis Varlamis (Harokopio University of Athens), Giuliana Vitiello (University of Salerno), Deqing Wang (Beihang University), Xin Xu (George Washington University), Linquan Yang (China University of Geosciences, Wuhan), Peng Zhang (Stony Brook University), Qian Zhang (Northeastern University), Xiang Zhao (National University of Defense Technology), Yishi Zhao (China University of Geosciences, Wuhan), and Zejun Zuo (China University of Geosciences, Wuhan). We would like to thank Dr. Sos Agaian from the City University of New York and Dr. Jinshan Tang from the Michigan Technological University for their inspiring keynote speeches on "Bioinspired Computational Urban Intelligence: A Long Road Ahead" and "Object Recognition Based on 3D Shapes Extracted from 3D Imaging and Applications," respectively. We are also very grateful to Aaron Schiller and Clement Wilson at Springer for their continuous support.

Contents

Part I
Technology and Infrastructure for Urban Intelligence

Controller Placements for Improving Flow Set-Up Reliability of Software-Defined Networks

Yuqi Fan, Tao Ouyang, and Xiaohui Yuan

Abstract Software-defined networking (SDN) is a new networking paradigm that decouples control plane from the data plane. A switch in the data plane device sends a flow set-up request to the controller, a device in the control plane, upon the arrival of an unknown flow. The controller responds the request with a flow entry to be installed in the flow table of the switch. Link failures can cause disconnections between switches and controllers. Most existing research on controller placement in SDNs investigated controller placements without considering single-link-failure impact on the number of dropped flow set-up requests in SDNs. In this paper, we formulate a novel SDN controller placement problem with the aim to minimize the average number of dropped flow set-up requests due to the single-link-failure. We propose two efficient algorithms for multiple-controller placements. The simulation results demonstrate that the proposed algorithms achieve competitive performance in terms of average number of dropped flow set-up requests under single-link-failure and average latency of flow set-up requests.

Keywords Software-defined network · Reliability · Single-link-failure · Network controller

1 Introduction

Software-defined networking (SDN) is a new networking paradigm that decouples control plane from data plane [1, 2]. Multiple-controller architectures have been introduced in SDNs and raised a new problem, the controller placement problem,

Y. Fan (✉) · T. Ouyang
School of Computer and Information, Hefei University of Technology, Hefei, Anhui, China
e-mail: yuqi.fan@hfut.edu.cn; ouyangtao@mail.hfut.edu.cn

X. Yuan
Department of Computer Science and Engineering, University of North Texas, Denton, TX, USA
e-mail: Xiaohui.Yuan@unt.edu

© Springer Nature Switzerland AG 2020
X. Yuan, M. Elhoseny (eds.), *Urban Intelligence and Applications*, Studies in Distributed Intelligence, https://doi.org/10.1007/978-3-030-45099-1_1

which needs to decide the controllers positions and how to associate switches with the controllers, since random placement is far from optimal [3, 4].

Some research has been conducted on the controller placement problem with the objective of minimizing the node-to-controller latency. The controller placement problem was first proposed by Heller et al. in [3] to minimize the communication latency between the switches and the controllers. A latency metric to minimize the total cost of flow set-up request from switches to controllers was introduced to deal with the mapping between the switches and the controllers under dynamic flow variations, and the metric considered the weight of switches and the delay from the switches to the controllers simultaneously, where the weight of a switch was related to the node degree of the switch and the maximum node degree in the network [5]. A network partition based scheme was designed, where the network was portioned into multiple subnetworks with revised k-means algorithm and a controller was placed in each subnetwork to minimize the maximum latency between the controller and the associated switches in the subnetwork [6]. A framework for deploying multiple controllers within a WAN was proposed to dynamically adjust the number of active controllers and delegate each controller with a subset of switches according to network dynamics [7].

The reliability is also an important performance metric for networks. A metric called expected percentage of control path loss due to failed network component was introduced to characterize the reliability of SDN networks, and a heuristic algorithm l-w-greedy was proposed to analyze the trade-off between reliability and latency; the expected percentage of control path loss was related to the number of control paths going through a component and the failure probability of the component [8, 9]. A controller placement strategy, Survivor, was proposed to explore the path diversity to optimize the survivability of networks with the aim to maximize the number of node-disjoint paths between the switches and the controllers; the strategy enhanced connectivity by explicitly considering path diversity, avoided controller overload by adding capacity-awareness in the controller placement, and improved failover mechanisms by means of a methodology for composing the list of backup paths [10]. The latency-aware reliable controller placement problem was investigated by jointly taking into account both the communication reliability and the communication latency between the controllers and the switches if any link in the network fails [11].

Link failures incur the breakdown of part of the network, during which some flow set-up requests from the switches are unable to reach the corresponding controllers and hence get dropped. To the best of our knowledge, very little attention in literature has ever been paid on the single-link-failure impact on the number of dropped flow set-up requests in SDNs. In this paper, we tackle the multiple-controller placement problem with the aim to improve the reliability in terms of the number of dropped flow set-up requests under single-link-failure.

The main contributions of this paper are as follows. We address the controller placement problem to maximize the reliability of the flow set-up requests under single-link-failure. We define a novel controller placement metric, the average number of dropped flow set-up requests, and propose two efficient algorithms for multiple-controller placements based on the proposed placement metric. We also evaluate the performance of the proposed algorithms through simulations. Experimental results demonstrate the proposed algorithms are very promising.

The rest of the paper is organized as follows. Section 2 presents the problem of this work. Section 3 discusses the proposed two algorithms: Reliability Aware Controller placement (RAC) and Fast-RAC (FRAC). Section 4 discusses our experimental evaluation. Section 5 concludes our work with a summary.

2 Problem Formulation

We model an SDN network topology as graph $G = (V, E)$, where V is the set of switches (or nodes) and E is the set of links. Each controller is co-located with a switch, and each switch is mapped to one controller. We assume that there is at most one link failure in the network [12]. The notations used in the paper are listed in Table 1.

When link e on the control path $p_{i,k}$ fails, we calculate the number of dropped flow set-up requests as follows:

$$\mathcal{D}(e) = \sum_{s_i \in V} \sum_{c_k \in C} r_{i,k} \cdot p_{i,k}^e \cdot x_{i,k}. \tag{1}$$

Our objective is to minimize the average number of dropped flow set-up requests by

Table 1 Symbols and notations used in our description

Notation	Description
s_i	Node/switch i
c_k	Controller c_k
C	Controller set
K	The number of controllers
N	The number of nodes/switches
L	The set of the links in all the control paths
u_k	The processing capacity of controller c_k
$r_{i,k}$	The number of requests from switch s_i to the mapped controller c_k
$x_{i,k}$	Indicate whether switch s_i is mapped to controller c_k ($= 1$) or not ($= 0$)
$y_{i,k}$	Denote whether controller c_k is co-located switch s_i ($= 1$) or not ($= 0$)
$p_{i,k}$	The link set on the control path between switch s_i and controller c_k
$p_{i,k}^e$	Denote whether link e is a link on control path $p_{i,k}$ ($= 1$) or not ($= 0$)

minimizing

$$\bar{\mathcal{D}} = \frac{\sum\limits_{e \in L} \mathcal{D}(e)}{|L|} \tag{2}$$

subject to

$$\sum_{k=1}^{K} x_{i,k} = 1, \quad \forall s_i \in V. \tag{3}$$

$$\sum_{i=1}^{N} y_{i,k} = 1, \quad \forall c_k \in C. \tag{4}$$

$$y_{i,k} \leq x_{i,k}, \quad \forall s_i \in V, \forall c_k \in C. \tag{5}$$

$$\sum_{i=1}^{N} x_{i,k} \cdot r_{i,k} \leq u_k, \quad \forall c_k \in C. \tag{6}$$

where $|L|$ denotes the number of links in L, and the average number of dropped flow set-up requests due to single-link-failure in the control paths are defined with Eq. (2). Equation (3) ensures that each switch is mapped to one and only one controller. Equation (4) mandates that each controller is placed onto exactly one switch. Equation (5) dictates that switch s_i is mapped to controller c_k if controller c_k is co-located with switch s_i. Equation (6) signifies that the number of requests to the controller cannot exceed the processing capacity of the controller.

3   RAC and FRAC Controller Placement Algorithms

In this section, we propose two controller placement algorithms, flow set-up request Reliability Aware Controller placement (RAC) and Fast-RAC (FRAC).

3.1   Reliability Aware Controller Placement Algorithm

Initially, algorithm RAC assumes that there are N controllers and each controller is co-located with a switch. The algorithm removes the redundant controllers iteratively until the number of controllers is K. For each controller, the algorithm evaluates the cost of removing it (steps 2–10). Assume the set of switches needing to be re-mapped after removing controller c_k is S_k. For each switch $s_i \in S_k$, algorithm RAC chooses the controller which incurs the least cost. During the re-mapping, Eq. (6) should be satisfied. After re-mapping all the switches in S_k, the algorithm can obtain the cost of removing c_k. Algorithm RAC evaluates the removal cost for all the controllers and removes the one which incurs the least cost (steps 11–12).

Algorithm 1 *RAC*

Input: Network topology $G = (V, E)$,
             The number of requests from the switches,
             The number of controllers K
Output: The set of locations placed with controllers C_p,
              Mapping relationship between switches and controllers
1:   Place a controller at the location of each switch, map each switch to the co-located controller,
     $C_p = V$, the number of placed controllers is $K' = N$;
2: **while** $K' \neq K$ **do**
3:     $\bar{\mathcal{D}}_{min} = \infty$;
4:     **for** each location $k \in C_p$ **do**
5:        Assume c_k is removed, and denote the set of switches mapped to c_k as S_k;
6:        Re-map each switch in S_k to the controller incurring the least average number of dropped
          flow set-up requests, and denote the average number of dropped flow set-up requests after
          re-mapping all the switches in S_k as $\bar{\mathcal{D}}_k$;
7:        **if** $\bar{\mathcal{D}}_k < \bar{\mathcal{D}}_{min}$ **then**
8:           $k_{min} = k, \bar{\mathcal{D}}_{min} = \bar{\mathcal{D}}_k$;
9:        **end if**
10:    **end for**
11:    Remove the controller $c_{k_{min}}$ which incurs the least average number of dropped flow set-up
      requests $\bar{\mathcal{D}}_{min}$;
12:    $C_p = C_p \setminus k_{min}, K' = K' - 1$;
13: **end while**

Time Complexity of Algorithm RAC The algorithm needs to remove $N - K$ controllers. In the worst case, a controller manages $N - K + 1$ switches. If the controller is removed, the algorithm performs re-mapping for the $O(N - K + 1)$ switches. The re-mapping for a switch checks at most $N - 1$ controllers. We can calculate all the shortest paths between all the node pairs in the network within $O(N^2 \cdot N) = O(N^3)$ time so that we can get the link set on the control path between each node pair before performing algorithm RAC, and hence the calculation of the average number of dropped flow set-up requests when the switch is mapped to the controller runs in time $O(N)$. Therefore, the time complexity of algorithm RAC is $O((N - K) \cdot N \cdot (N - K + 1) \cdot (N - 1) \cdot N) = O(N^3 \cdot (N - K)^2) = O(N^5)$, since $K \ll N$.

3.2 Fast-RAC

We propose algorithm Fast-RAC (FRAC) to reduce the time complexity of algorithm RAC. FRAC maintains a mapping controller priority list PL_i for each switch s_i, where each item in the list is a controller and the controllers in the list are sorted in the non-ascending order of the path lengths between the switch and the controllers. FRAC uses two arrays *Current* and *Next*, each with length N. *Current*$[i] = k$ denotes that switch s_i is currently mapped to controller c_k. Initially, each switch is mapped to the co-located controller, that is, *Current*$[i] = i$. *Next*$[i] = k'$ indicates that switch s_i will be mapped to controller $c_{k'}$ if controller c_k is removed, where $c_{k'}$ will cause the least number of dropped flow set-up requests caused by single-link-failure. We initialize arrays *Current* and *Next* before placing a controller at each switch (step 1), and update them when a controller is removed (step 11). During the update, if switch s_i is re-mapped from c_k to $c_{k'}$, $Current[i] = k$ is updated as $Current[i] = k'$ and $Next[i]$ should be updated by searching the first controller available in the list PL_i.

Time Complexity of Algorithm FRAC The construction of the mapping controller priority lists for all the switches can be performed in time $O(N^3)$ by calculating the shortest paths between all the node pairs. The initialization and update of the array *Current* can be performed in time $O(N)$. Array *Next* can also be constructed in time $O(N)$, while the update of the array *Next* is conducted in time $O(N^2)$, since FRAC checks at most $N - 2$ controllers to find the controllers available for each of the N switches. The time consumed to re-map a switch is reduced from $O(N)$ with RAC to $O(1)$ with FRAC. Therefore, the time complexity of algorithm FRAC is $O(N^4)$.

4   Performance Evaluation

In this section, we evaluate the performance of the proposed controller placement algorithm. We also investigate the impact of important parameters on the performance of the proposed algorithms.

4.1   Simulation Set-up

We compare the proposed algorithms RAC, FRAC against the state of the arts: l-w-greedy [9], SVVR [10], and CPP [3]. The two coefficients l and w are set as $l = 2$ and $w = 1$ to enable l-w-greedy to achieve the best performance as described in [9]. The network topologies used in the simulation are ATT (ATT North America) and Internet2 (Internet2 OS3E) [13, 14]. All the controllers have the same processing capacity of 1800 kilorequests/s [10]. The requests from the switches are generated with uniform distribution pattern. 30 set of requests are generated for each request distribution pattern randomly, while the average number of flow set-up requests of the switches are 200 kilorequests/s. We use geographical distance as an approximation for latency [3].

4.2   Evaluation Metrics

The performance metrics evaluated are as follows:

1. The average number of dropped flow set-up requests under single-link-failure as calculated via Eq. (2).
2. The average latency of flow set-up requests $\bar{l}$, defined via Eq. (7), where $l_{i,k}$ denotes the communication latency of control path $p_{i,k}$.

$$\bar{l} = \frac{\sum\limits_{s_i \in V} \sum\limits_{c_k \in C} r_{i,k} \cdot x_{i,k} \cdot l_{i,k}}{\sum\limits_{s_i \in V} \sum\limits_{c_k \in C} r_{i,k} \cdot x_{i,k}}. \tag{7}$$

4.3 Simulation Results

4.3.1 Average Number of Dropped Flow Set-Up Requests

In this section, we evaluate the average number of dropped flow set-up requests of each algorithm by varying the number of controllers from 5 to 10.

Figure 1 shows that both algorithm RAC and algorithm FRAC obtain better performance than the benchmark algorithms because the proposed algorithms minimize the average number of dropped flow set-up requests when placing the controllers in the network. Algorithm RAC performs slightly better than FRAC since RAC removes the controller which incurs the least average number of dropped flow set-up requests, while FRAC finds the controller leading to the least number of dropped flow set-up requests.

Algorithm CPP achieves the best performance among three benchmark algorithms. Algorithm SVVR aims to maximize the number of disjoint paths between switches and controllers, l-w-greedy tries to minimize the expected percentage of control path loss, and CPP deploys the controllers with the objective of optimizing the average communication latency of switches and controllers. Algorithm CPP potentially aggregates the requests on a subset of the network, which reduces the number of links on the control paths and the total number of requests caused by the single-link-failure.

4.3.2 Average Latency of Flow Set-Up Requests

In this section, we evaluate the average latency of flow set-up requests of each algorithm, assuming the number of controllers varies from 5 to 10.

Figure 2 depicts the average latency of flow set-up requests of RAC, FRAC, SVVR, l-w-greedy, and CPP. Algorithms RAC and FRAC achieve a similar performance. Algorithm l-w-greedy results in the worst performance, because algorithm l-w-greedy aims to optimize the reliability of control path between switches and controllers, which potentially leads to long control paths between the switches and controllers. Algorithms RAC and FRAC place the controllers considering

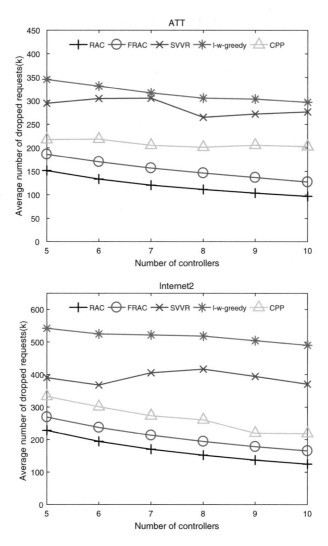

Fig. 1 The average number of dropped requests under different number of controllers

the number of dropped flow set-up requests under single-link-failure, so the two algorithms put the controllers close to the switches which generate a large number of flow set-up requests. Algorithm CPP minimizes the latency between the switches and the controllers without considering the number of flow set-up requests generated by the switches. Therefore, algorithm CPP obtains better performance than the other algorithms, where the distance between the switches and the controllers has an important impact on the controller placement.

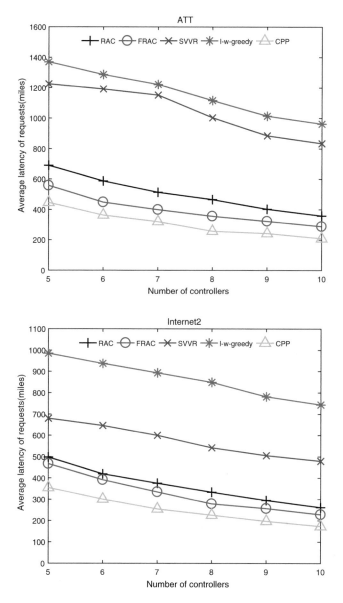

Fig. 2 The average latency of flow set-up requests under different number of controllers

5 Conclusions

Reliability is an important concern during controller placements in SDNs since link failures can cause disconnections between switches and controllers, and even incur cascading failures of other controllers. In this paper, we take the transmission reliability of flow set-up requests into consideration during controller placements. We formulated a novel SDN controller placement problem with the aim to minimize the average number of dropped flow set-up requests due to the single-link-failure. We propose flow set-up request Reliability Aware Controller placement (RAC) algorithm and Fast-RAC (FRAC) algorithm for multiple-controller placements. Experiments are conducted through simulations. Our experimental results demonstrate that the proposed algorithms achieve competitive performance in terms of average number of dropped flow set-up requests under single-link-failure and average latency of flow set-up requests.

Acknowledgments This work is partly supported by the National Natural Science Foundation of China (61701162, U1836102), the Anhui Provincial Natural Science Foundation (1608085MF142), and the open project of State Key Laboratory of Complex Electromagnetic Environment Effects on Electronics and Information System (CEMEE2018Z0102B).

References

1. B.A.A. Nunes, M. Mendonca, X.-N. Nguyen, K. Obraczka, T. Turletti, A survey of software-defined networking: past, present, and future of programmable networks. IEEE Commun. Surv. Tutor. **16**(3), 1617–1634 (2014)
2. M. Elhoseny, H. Elminir, A.M. Riad, X. Yuan, Recent advances of secure clustering protocols in wireless sensor networks. Int. J. Comput. Netw. Commun. Secur. **2**(11), 400–413 (2014)
3. B. Heller, R. Sherwood, N. McKeown, The controller placement problem, in *Proceedings of the First Workshop on Hot Topics in Software Defined Networks* (ACM, New York, 2012), pp. 7–12
4. X. Yuan, M. Elhoseny, H.K. El-Minir, A.M. Riad, A genetic algorithm-based, dynamic clustering method towards improved WSN longevity. J. Netw. Syst. Manage. **25**(1), 21–46 (2017)
5. L. Yao, P. Hong, W. Zhang, J. Li, D. Ni, Controller placement and flow based dynamic management problem towards SDN, in *2015 IEEE International Conference on Communication Workshop (ICCW)* (IEEE, New York, 2015), pp. 363–368
6. G. Wang, Y. Zhao, J. Huang, Q. Duan, J. Li, A k-means-based network partition algorithm for controller placement in software defined network, in *2016 IEEE International Conference on Communications (ICC)* (IEEE, New York, 2016), pp. 1–6
7. M.F. Bari, A.R. Roy, S.R. Chowdhury, Q. Zhang, M.F. Zhani, R. Ahmed, R. Boutaba, Dynamic controller provisioning in software defined networks, in *2013 9th International Conference on Network and Service Management (CNSM)* (IEEE, New York, 2013), pp. 18–25
8. Y. Hu, W. Wang, X. Gong, X. Que, S. Cheng, Reliability-aware controller placement for software-defined networks, in *Proceedings of the 2013 IFIP/IEEE International Symposium on Integrated Network Management (IM 2013)* (2013), pp. 672–675
9. Y. Hu, W. Wang, X. Gong, X. Que, S. Cheng, On reliability-optimized controller placement for software-defined networks. China Commun. **11**(2), 38–54 (2014)
10. L.F. Müller, R.R. Oliveira, M.C. Luizelli, L.P. Gaspary, M.P. Barcellos, Survivor: an enhanced controller placement strategy for improving SDN survivability, in *Global Communications Conference (GLOBECOM)* (IEEE, New York, 2014), pp. 1909–1915
11. Y. Fan, Y. Xia, W. Liang, X. Zhang, Latency-aware reliable controller placements in SDNs, in *International Conference on Communications and Networking in China* (Springer, Basel, 2016), pp. 152–162

12. A. Markopoulou, G. Iannaccone, S. Bhattacharyya, C.-N. Chuah, C. Diot, Characterization of failures in an IP backbone, in *INFOCOM 2004* (IEEE, New York, 2004), pp. 2307–2317
13. S. Knight, H.X. Nguyen, N. Falkner, R. Bowden, M. Roughan, The internet topology zoo. IEEE J. Sel. Areas Commun. **29**(9), 1765–1775 (2011)
14. Internet2 Open science, scholarship and services exchange. [Online]. Available: http://www.internet2.edu/network/ose/

A Multiple Compatible Compression Scheme Based on Tri-state Signal

Tian Chen, Yongsheng Zuo, Xin An, and Fuji Ren

Abstract A novel test data compression scheme based on the tri-state signal is proposed to solve the problem of increasing embedded chip test data in the development of Smart City. Firstly, partial inputs reduction is performed on the test set, and the don't care bits proportion is improved by merging the test patterns with a high bit ratio so that the compatibility of each test pattern is improved. After using the inputs reduction technology, using the characteristics of the tri-state signal, the test set is divided into several sub-segments and uses the tri-state signal to compress the sub-segments with compatible coding, and the compression rate of the test set is improved by considering multiple compatible rules. The experimental results show that compared with the previous work results, the proposed scheme achieves a good compression ratio, the average test compression ratio can reach 82.15%. At the same time, the test power and area overhead are not significantly improved.

Keywords Compression · Automatic test equipment · Circuit

T. Chen (✉) · Y. Zuo · X. An
School of Computer and Information, Hefei University of Technology, Hefei, Anhui, China

Anhui Province Key Laboratory of Affective Computing and Advanced Intelligent Machine, Hefei University of Technology, Hefei, Anhui, China
e-mail: ct@hfut.edu.cn

F. Ren
School of Computer Science and Information Engineering, Hefei University of Technology, Hefei, Anhui, China

Anhui Province Key Laboratory of Affective Computing and Advanced Intelligent Machine, Hefei University of Technology, Hefei, Anhui, China

Faculty of Engineering, The University of Tokushima, Tokushima, Japan

© Springer Nature Switzerland AG 2020
X. Yuan, M. Elhoseny (eds.), *Urban Intelligence and Applications*, Studies in Distributed Intelligence, https://doi.org/10.1007/978-3-030-45099-1_2

1   Introduction

With the continuous development of information technology, the level of application of urban information technology has been continuously improved, and the construction of the smart cities has emerged. As a deep expansion and integration application of information technology, smart city is one of the important directions for the breakthrough of the new generation of information technology and an important part of the development of global strategic emerging industries. Building a smart city is of great significance in realizing sustainable urban development, leading the application of information technology, improving the overall competitiveness of cities. In the process of building a smart city, the development of the Internet of Things and the smart city is inseparable, and the embedded integrated circuit chip in the Internet of Things, with the development of smart cities, the number of transistors in the chip is constantly increasing, in order to ensure the failure of the embedded chip. Coverage and test data volume have also increased correspondingly, which poses certain challenges to the storage capacity and bandwidth of automatic test equipment (ATE). More test data means an increase in test time and test power consumption. The cost of using ATE, the test time, and the increase in the amount of test data are the main factors that increase the overall test cost.

Test data compression technology can effectively cope with the increase of test data [1]. The code-based compression method is a very important type of test data compression technology, and this method can be well compatible with the design process. The method first encodes the pre-computed test set into a smaller data set, and then stores the compressed data set in the ATE. During the test, the compressed set enters the chip through the communication channel between the ATE and the chip, and then the decompressed structure generates test patterns that are compatible with the original test data. Typically, a pre-computed test set typically includes "0," "1," and don't care bits (X), referred to as a test cube. The test cube is encode into coded data which represented by logical values "0" and "1." On the chip, the test pattern generated by the decompressed structure is compatible with the original test cube, but only contains binary data streams of "0" and "1." However, the ATE supports not only logical values "0" and "1," but also high-impedance signals (hi-Z) [2]. This provides a new idea for the code-based compression method. The test cubes can be encoded with tri-state signals ("0", "1" and "hi-Z"), and the compressed set containing the tri-state signals can be stored on the ATE. The information stored in this way only needs to use a structure which can convert the tri-state signal into binary logic on the chip, then combines with the appropriate decompression structure to realize a code-based method based on the tri-state signal.

This paper designs a test data compression scheme based on tri-state signal coding. The tri-state signal refers to logic values "0," "1," and hi-Z signal. The tri-state signal has one more bit than the binary signal so that the amount of information carried by the same length code can be increased during encoding. In order to obtain a good compression ratio, firstly, the scheme in this paper compresses the test cube by inputs reduction compression, then multiple-compatibility coding compression is used on the test cube according to the tri-state signal. The rest of this article is organized as follows. Section 2 describes the tri-state signal, partial inputs reduction technology, and other related work. In Sect. 3, this paper describes the entire process and coding scheme for test data compression. This paper also provides an example to describe the encoding process. Section 4 describes a decompression structure consisting of a tri-state signal conversion circuit and a decompression circuit. This paper shows the experimental results in Sect. 5. Finally, it summarizes full text in Sect. 6.

2 Preliminaries

2.1 Tri-state Signal Detection Circuit

Test cubes consist of 0s,1s, and Xs. But in the traditional methods, when the test cube is stored in ATE or delivered to the chip, all Xs should be filled with 0s or 1s . Now, the tri-state can be used in storage and transmission. Most of the internal logic circuits cannot recognize the high-impedance signal, the problem is how to detect the tri-state signal on the chip and reconvert the input signal data containing the tri-state into test data containing only "0" and "1." In 1993, the patent [3] proposed a tri-state signal detection circuit that provides a specific pin and detector for detecting a tri-state signal. The input port connected to the pin can detect a logic level of "0," "1," or high-impedance signal. The detector has a register for storing the logic level of the pin. The decoder determines the state of the pin from the logic level stored in the register, so that the tri-state signal can be detected, but the hardware overhead of the circuit is very huge. Subsequently, in 2001, the patent [4] propose another relatively simple tri-state signal detection circuit. This tri-state input detection circuit is shown in Fig. 1. The circuit has a tri-state signal input port and two binary output port. This circuit can generate a unique binary output sequence for each of the input signal states through the internal six transistors. The unique binary output sequence is used to mark the tri-state input signal. At the same

Fig. 1 Tri-state detector circuit in [4]

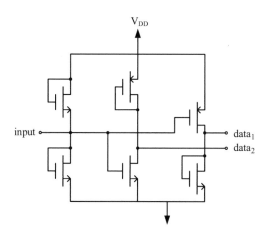

Table 1 Truth table of tri-state signals detection circuit

	Detection output	
Input	$data_1$	$data_2$
0	1	1
1	0	0
Hi-Z	1	0

time, the entire tri-state detection circuit contains only six transistors, and the area overhead and complexity is small, and the binary output generated by the circuit can be easily combined with the control module of the decompression circuit, so this tri-state detection is selected in this paper. The circuit is used to decompress the structure and combine the other logic gates to apply the tri-state signal to the decompression of the test data. The truth table of the tri-state signal is shown in Table 1.

2.2 Partial Inputs Reduction Technology

In order to reduce the number of bits in the test cube and increase the probability of compatibility between test patterns, the scheme of this paper needs to use inputs reduction technology to handle the test cube before test data compression. Inputs reduction technology [5] is a technique for reducing the number of test cube by reducing the number of bits which are compatible or reverse compatible. The traditional inputs reduction technology reduces the number of all compatible or reverses compatible columns in the test cube. As the number of reduced bits increases, the difficulty of routing increases, and the test power during testing the difficulty of routing and the test power is increased. Partial inputs reduction technology is a compromise between the compression ratio and the difficulty of routing. The method is to reduce the number of columns with a high bit rate so that a large number of don't care bits can be reserved for subsequent test compression. Figure 2 shows an example of traditional full inputs reduction and partial inputs reduction compression for a test cube. This paper will use partial inputs reduction compression. After the partial inputs reduction, the number of specified bits are reduced. This method makes the test cube more conducive to data compression.

3   Compression Scheme

3.1   Test Data Compression Process

The encoding scheme of this paper first divides the pre-computed test patterns of the determined test cube into fixed-length data blocks, then encodes the data blocks. Use a shorter new codeword to replace the original data block to achieve the purpose of compressing the data length. Figure 3 is the overall test flow of the solution in this

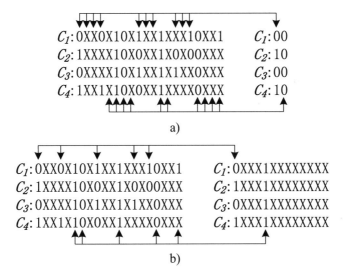

Fig. 2 An example of inputs reduction. (**a**) Traditional input reduction. (**b**) Partial input reduction

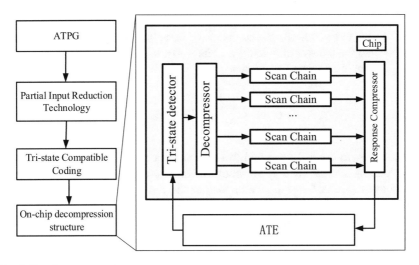

Fig. 3 Test data compression process

paper. The test cube is first generated by automatic test pattern generation (ATPG), then the test cube is processed by partial inputs reduction technology; finally, used a tri-state compatible coding scheme for the test cube and the encoded data is stored in the ATE. In the test, ATE transfers the encoded data to the chip, recognizes and converts the tri-state signal into binary data through the tri-state detection circuit, then decodes the encoded data through the decompressor and sends the data to the scan chain.

3.2 Compatible Coding Theory

The following is a theoretical analysis of the mathematical expectation of the number of bits that need to be jumped from one test mode to another and related to which factors. Suppose that the test cube is determined to be T, the cardinality of the cube is $t = |T|$. Each test cube has a length of n, and D is a test cube containing s determined bits, and the numbers of no care bits in D are $n - s$. Assuming that D_m is a test mode of length n, the probability that D_m and D have the largest m ($m - s$) bits collision is

$$
P_m = \begin{cases} \frac{t \cdot d_m}{2^n}, & d_m = 2^{n-s} \sum_{m}^{i=0} \binom{i}{s}(t \cdot d_m < 2^n); \\ 1, & (t \cdot d_m \geq 2^n). \end{cases}
\tag{1}
$$

where d_m represents the total number of test patterns that have a maximum of m bits in conflict with D. Simplify the formula (1) to get the formula (2):

$$
P_m = \begin{cases} \frac{t}{2^s} \sum_{m}^{i=0} \binom{i}{s}, & (\sum_{m}^{i=0} \binom{i}{s} < \frac{2^s}{t}); \\ 1, & (\sum_{m}^{i=0} \binom{i}{s} \geq \frac{2^s}{t}). \end{cases}
\tag{2}
$$

According to formula (2), the number m of jump bits depends on the size t of the test cube and the number of determined bits s. The mathematical expectation of the jump position is calculated as follows:

$$
E(t, s) = \sum_{s}^{m=1} t \cdot (P_m - P_{m-1}).
\tag{3}
$$

According to Table 2, assuming that the test cube is 1000 and the test cube has a certain number of bits of about 20, then any test pattern can jump to a certain test mode in the test cube as long as the jump is 2.78 bits (Table 2). The mathematical expectation is that an arbitrary test pattern and test cube determine the average minimum Hamming distance of the test cube. In fact, the test mode in which the jump occurs also belongs to the same determined test cube. For a certain test cube, the determined bits of the test cube are often associated with the same circuit structure and therefore have a certain bit correlation. According to the above theoretical basis, the value of the transition between each test mode in the test cube

Table 2 Mathematical expectation of the number of hopping bits

s/t	5	10	20	30	40
10	0.69	0.43	6.14	–	–
100	0.00	0.90	4.26	7.84	–
1000	0.00	0.02	2.78	6.09	9.54
10,000	0.00	0.00	1.79	4.66	7.83

Table 3 Coding scheme table

Coding data	Prefix	Suffix	Codeword
All bits are 0 Except X	0	Z	0Z
All bits are 1 Except X	1	Z	1Z
Forward partially compatible	Z	Z	Z1–Z/ZZ
Reverse shift compatible	Z0	Z	Z0–Z/Z0Z

is actually small and has correlation, so the test cube can be encoded by using the characteristic of the Hamming distance.

3.3   Based on Multiple Compatible Tri-state Coding Scheme

The coding scheme in this paper adopts a variable length compatible coding method as shown in Table 3. For a test cube $T:\{C_1, C_2, \ldots, C_N\}$ with N test patterns of length l, each pattern C_i is divided into m sub-patterns with the same length according to the length d of the multi-scan chain. The sub-patterns are called $M:\{S_{i1}, S_{i2}, \ldots, S_{iN}\}$. And when the sub-pattern's length less than d, the sub-pattern is completed with don't care bits. Each pattern $S_{i,(j+1)}$ is coded with reference to the previous pattern $S_{i,j}$. The initial reference pattern goes to all zero. According to the patterns $S_{i,j}$, the next pattern $S_{i,(j+1)}$ is divided into four cases:

1. All bits are "0" except don't care bits. It is represented by the prefix "0" and $S_{i,(j+1)}$ is filled with all "0." Expressed by the prefix "0," it is filled with all "0," the suffix is "Z," so the code word is "0Z."
2. All bits are "1" except don't care bits. It is represented by the prefix "1" and $S_{i,(j+1)}$ is filled with all 1. Expressed by the prefix "1," it is filled with all "1," the suffix is "Z," so the code word is "1Z."
3. For the two test patterns C_i and C_j, if there exist k bits at most that are forward-compatible from the first bit, then these two test patterns C_i and C_j are positively compatible with the k-bit portion. Therefore, the encoding method is represented by the prefix "Z," and the full forward compatibility is regarded as a special case of partial forward compatibility. The forward compatibility is all compatible, the encoding is "ZZ"; for other non-special cases, the former is the d bits are all incompatible; the don't care bits of $S_{i,(j+1)}$ are filled in as the corresponding position don't care bits of $S_{i,j}$. In order to be able to calculate $S_{i,(j+1)}$ from $S_{i,j}$, the mask of the incompatible part of $S_{i,j}$ and $S_{i,(j+1)}$ need to be marked in the codeword.
4. If the pattern S_i obtained by shifting the right bit of the pattern C_i to the right is compatible with C_i, the pattern C_i and C_j are said to be shift compatible. Inverse shift compatible with the reference pattern $S_{i,j}$. The prefix "Z0" means that because the forward part is compatible, the first two bits of the code word must be "Z1" or "ZZ." The unspecified bits in pattern $S_{i,(j+1)}$ are rotated to the right of the corresponding bit position after the bit shift, The unspecified bits in

Table 4 An example of compression

Set	Subsets	Reference pattern	Encoded	Test data status	Bits
	S_{11}:X00X1X1X	00000000	Z1010Z	Partially compatible	6
S1	S_{12}:XX0X10XX	00001010	ZZ	Froward compatible	2
	S_{13}:X11X0X0X	00001010	Z0Z	Reverse compatible	3
	S_{21}:1XXX0X1X	11110101	Z01Z	Reverse shift Compatible	4
S2	S_{22}:XXXXXXXX	10000010	0Z	All X(0)	2
	S_{23}:X11X1X1X	00000000	1Z	All X(1)	2
Total	48 bits	–	–	–	19 bits

$S_{i,(j+1)}$ are filled with the specified bits of the $S_{i,j}$ corresponding positions after cyclic right shift by q bits, the code of $q-1$ is marked in the code word to restore the pattern $S_{i,(j+1)}$.Each codeword is represented by the suffix "Z."

When coding, all "0" and all "1" are coded preferentially, the test pattern will be encoded to all "0" or all "1" according to the higher bits of the last sub-pattern. The purpose is to reduce the probability of the scan chain's rotation and reduce the shift power.

3.4 Coding Example

Table 4 shows an example of coding a test cube. The original test patterns S_1 and S_2 have 48 bits, the test patterns are divided into six sub-patterns. After coding, the test cube is compressed to 19 bits.

4 Decompressor

4.1 Tri-state Conversion Circuit

After the tri-state detection circuit, the single input signal from the ATE becomes two output signals. When the input signal carries logical values "0" and "1," the logical values of the two output signals are not directly available for decompression of the subsequent test data. Therefore, this paper adds a conversion circuit between the tri-state detection circuit and the decompression structure. The conversion circuit is shown in Fig. 4. The original output values $data_1$ and $data_2$ are converted to $data'_1$ and $data'_2$ after passing through two logic gates. When the input signal is "0" or "1," $data'_2$ is "0," indicating that the input signal is a logical value, and the real logical value is $data'_1$. When the input signal is in a high-impedance state, $data'_2$ is "1" and it is easy to distinguish. The truth table of tri-state signal is shown in Table 5.

Fig. 4 Tri-state conversion circuit

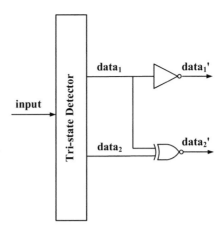

Table 5 Truth table of tri-state signal

Input	Detection output		Conversion output	
	$data_1$	$data_2$	$data_1'$	$data_2'$
0	1	1	0	1
1	0	0	1	1
Hi-Z	1	0	0	0

Figure 5 shows the decompression structure of the encoding scheme in this paper. The decompression structure includes a tri-state signal detection circuit, a control signal generator (CRU), two settable counters, a decoder, a readable and writable register, one plus twisted-loop counters, several logic gates for XOR gates and multiplexers. Where d is the length of the scan chain and m is the number of scan chains. The CRU is used to analyze the decoding type and issue signal control twisted-loop counters, multiplexers, logic gates, etc. for decoding, and outputs the decoded test patterns to different scan chains. The register is used to store partially compatible data temporarily. The shift value when the mask is compatible is the same with the reverse shift.

5 Experimental Results

In order to verify the effectiveness of the compression scheme in this paper, this paper selects some circuits in the ISCAS'89 to perform experiments [6]. Table 6 shows the experimental circuit information after partial inputs reduction.

In the scheme of this paper, the segment length d of the test pattern has an important influence on the compression efficiency of the test data. Figure 6 shows the compression ratio change of the compression scheme in this paper under different segment lengths. When the segment length is too short, the flag bit occupies a relatively high proportion in the entire code words, and it is difficult to obtain a high compression ratio; when the segmentation length is too long, the compatibility

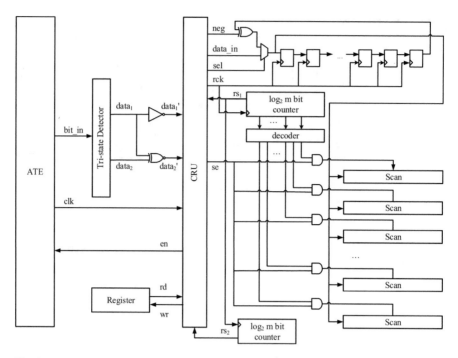

Fig. 5 Decompression structure

Table 6 The details of mintest [6] test sets

Circuit	Scanning cells	Test pattern	Data volume	Rate of X
s5378	197	111	23,754	70.8
s9234	247	159	39,273	70.4
s13207	620	236	165,200	92.4
s15850	538	126	76,986	81.7
s38417	1572	99	164,736	66.6
s38584	1408	136	199,104	81.7

probability between the sub-pattern is greatly reduced, which is very disadvantageous to the test data compression. Even if the lengths do not differ greatly, the compression rate fluctuates in a certain range. Therefore, choosing a suitable segment length will greatly help improve the compression ratio of the test data.

Table 7 shows the comparison of the compression ratios of this paper's scheme with other compression schemes. The last column is the compression ratio of the optimal scan chain length was chosen in the scheme of this paper. The results show that the compression ratio of the scheme in this paper is higher than other schemes. The reason why the scheme in this paper can achieve a good compression ratio is that: use partial inputs reduction for the test cube before compression; after this process, the test cube is more beneficial for test data compression; when performing

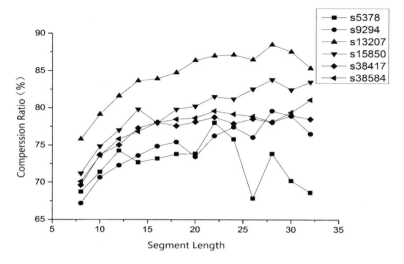

Fig. 6 Compression ratio under different segment length

Table 7 The comparison of compression ratio with variety of previous schemes (%)

Circuit	Golomb [7]	FDR [8]	VIHC [9]	EFDR [10]	LPMC [11]	MRSR [12]	TSC [2]	Proposed
s5378	37.11	47.98	51.78	51.93	57.76	60.41	75.39	78.05
s9234	45.35	43.61	47.25	45.89	56.21	60.45	72.49	79.61
s13207	79.74	81.30	83.51	81.85	86.32	87.66	87.39	89.02
s15850	62.82	66.21	67.94	67.99	69.68	72.50	79.11	85.12
s38417	28.37	43.37	53.36	60.57	65.98	60.65	76.65	79.97
s38584	57.17	60.93	62.28	62.91	68.68	73.15	79.72	81.11
Avg.	51.74	57.23	60.29	61.86	67.27	69.14	78.64	82.15

tri-state signal encoding, it is not only considered that the test patterns are directly compatible but also the case of partial compatibility and shift compatibility.

Table 8 compares the area overhead of different scenarios. This paper uses Synopsys' DC tools to synthesize and analyze the decompressed structure and the corresponding reference circuit. The area overhead calculation formula:

$$\text{Area Overhead} = \frac{\text{Decompression Structure Area}}{\text{Reference Circuit Area}} * 100\% \tag{4}$$

Although this paper uses a tri-state detection circuit, it consists of only six transistors, and the area overhead is negligible. By comparison, the area overhead of the decoding circuit is approximately between the LMPC encoding and the TSC encoding, and there is no significant increase in comparison with other schemes. However, the scheme in this paper has a significant increase in compression ratio over several schemes. It is acceptable to reduce the cost of chip testing by increasing the compression ratio in terms of area overhead.

Table 8 Comparison of the hardware area overhead with variety of previous test data compression (%)

Circuit	Golomb [7]	FDR [8]	VIHC [9]	EFDR [10]	LPMC [11]	TSC [2]	Proposed
s5378	4.0	7.8	5.8	8.3	10.3	8.1	10.1
s9234	3.2	5.9	4.6	6.3	8.1	6.3	9.2
s13207	4.1	3.5	2.2	3.7	4.6	2.8	4.5
s15850	2.0	3.6	2.3	3.8	4.7	2.7	3.9
s38417	0.5	1.4	0.7	1.5	1.9	1.2	1.5
s38584	0.7	1.5	0.7	1.6	2.0	1.0	1.6

Table 9 Comparison of shift power

Circuit	Mintest [6]	OC_SP [13]	FDR [8]	ERLC [14]	LPMC [11]	Proposed
s9234	1463	8132	5692	3500	3699	4058
s13207	122,031	17,809	12,416	8155	7854	9956
s15850	90,899	24,850	20,742	13,450	12,980	15,781
s38417	601,840	578,450	172,665	120,775	115,682	130,045
s38584	535,875	108,050	136,634	89,356	88,632	110,245
Avg.	273,055	147,458	69,630	47,039	45,769	54,071

Table 9 shows the comparison results of shift power consumption between this scheme and other methods on different circuits. The shift power is calculated using the Weighted Transition Metric (WTM) [15]:

$$\text{WTM}_i = \sum_{l-1}^{j=1} [(b_{i,j} \oplus b_{i,j+1}) \times (l - j)], \tag{5}$$

where l is the length of the scan chain and $b_{i,j}$ is the value of the j-th bit in the test pattern that is sent to the scan chain at i-th bit. The value is "0" or "1." The results in Table 9 show that the shift power consumption of this method is slightly higher than that of the ERLC and the LMPC, the shift power consumption is lower than that of other methods.

6 Conclusion

The large amount of test data and high test cost are important issues that must be faced in chip testing. the problem of testing data volume increased, this paper proposes a multistage data compression method combining partial inputs reduction technology and compatible compression based on the tri-state signals for encoding. The method firstly uses partial inputs reduction technology to reduce the number of specified bits in the test cube, then the processed test cube was encoded by tri-state signal. The encoding not only considers the full compatibility but also considers the partial compatibility and shift compatibility; finally, the decompression structure of compressed data on-chip was designed. Experimental results show that the method in this paper achieves a good compression rate without excessive area overhead and lower test application time. At the same time, when the method is also compressed and filled, it controls the shift power in an acceptable range.

Acknowledgments This work sponsored by: The Key Program of the National Natural Science Foundation of China (Grant No. 61432004); The National Natural Science Foundation of China (Grant No.61474035, No.61204046, No.61502140); NSFC-Shenzhen Joint Foundation (Key Project) (Grant No.U1613217).

References

1. S. Mirthulla, A. Arulmurugan, Improvement of test data compression using combined encoding, in *International Conference on Electronics and Communication Systems (ICECS)* (2015), pp. 635–638
2. S. Seo, Y. Lee, S. Kang, Tri-state coding using reconfiguration of twisted ring counter for test data compression. IEEE Trans. Comput.-Aided Des. Integr. Circ. Syst. **35**(2), 274–284 (2016)
3. J. Nicolai, Integrated circuit with mode detection pin for tristate level detection, U.S. Patent 5198707 (1993)
4. D. Thomson, P. Sheridan, J. Cleary, Tri-state input detection circuit. U.S. Patent: 6133753 (2000)
5. C.A. Chen, S.K. Gupta, Efficient BIST TPG design and test cube compaction via input reduction. IEEE Trans. Comput.-Aided Des. Integr. Circ. Syst. **17**(8), 692–705 (2002)
6. K.M. Butler, J. Saxena, A. Jain, T. Fryars, Minimizing power consumption in scan testing: pattern generation and DFT techniques, in *International Conference on Test* (2004), pp.355–364
7. A. Chandra, K. Chakrabarty, System-on-a-chip test-data compression and decompression architectures based on Golomb codes. IEEE Trans. Comput.-Aided Des. Integr. Circ. Syst. **20**(3), 335–368 (2001)
8. A. Chandra, K. Chakrabarty, Test data compression and test resource partitioning for system-on-a-chip using frequency-directed run-length (FDR) codes. IEEE Trans. Comput. **52**(8), 1076–1088 (2003)
9. P.T. Gonciari, B.M. Al-Hashimi, Variable-length input Huffman coding for system-on-a-chip test. IEEE Trans. Comput.-Aided Des. Integr. Circ. Syst. **22**(6), 783–796 (2003)
10. A.H. El-Maleh, Test data compression for system-on-a-chip using extended frequency-directed run-length (EFDR) code. IET Comput. Digit. Tech. **2**(3), 155–163 (2008)
11. T. Chen, X. Yi, W. Wang, J. Liu, H. Liang, F. Ren, Low power multistage test data compression scheme. Acta Electron. Sin. **45**(6), 1384–1387 (2017)
12. K. Ji-shun, L. Jie-tang, Z. Liang, Test data compression method for multiple scan chain based on mirror-symmetrical reference slices. J. Electron. Inf. Technol. **37**(6), 1514–1518 (2015)
13. K.A. Bhavsar, U.S. Mehta, Analysis of don't care bits filling techniques for optimization of compression and scan power. Int. J. Comput. Appl. **18**(3), 887–975 (2011)
14. W. Zhan, A. EL-Maleh, A new collaborative scheme of test vector compression based on equal-run-length coding (ERLC), in *International Conference on Computer Supported Cooperative Work in Design* (2009), pp. 21–25
15. R. Sankaralingam, R.R. Oruganti, N.A. Touba, Static compaction techniques to control scan vector power dissipation, in *Proceedings of IEEE VLSI Test Symposium (VTS)* (2000), pp. 35–40

A Thin Client Error-Correcting Data Storage Framework Based on Blockchain

Yuqi Fan, JingLin Zou, Siyu Liu, Qiran Yin, Xin Guan, and Xiaohui Yuan

Abstract Traditional centralized and distributed data storage systems are hard to achieve data integrity. A blockchain is more resistant to modification of the data than widely used technologies, such as digital signature and digital watermarking. Each node in a blockchain is required to store data so that all the nodes should have powerful computing and storage capacity. Centralized service points make a thin client possible at the cost of decreasing decentralization that a blockchain possesses. In addition, a wrong block breaks the data consistency in the system, and existing data dissemination protocol cannot solve the wrong block problem efficiently. This paper proposes a thin client error-correcting data storage framework based on blockchain. The data of the clients are stored as blocks in the system to achieve the data integrity. The data requests are handled by randomly selected blockchain nodes. We propose two error-correcting mechanisms to validate and correct the wrong data blocks. We also analyze the performance of the proposed framework with respect to data tamper proof, consensus, and thin client. The proposed framework has the characteristics of lightweight client, data tamper proof, and high resource utilization in the blockchain network.

Keywords Blockchain · Data storage · Data tamper · Thin client

1 Introduction

Traditional centralized and distributed storage systems face the problem that data are prone to be tampered by outside or internal attackers. A blockchain is more resistant to modification of the data than widely used technologies, such as digital

Y. Fan (✉) · J. Zou · S. Liu · Q. Yin · X. Guan
School of Computer and Information, Hefei University of Technology, Hefei, Anhui, China
e-mail: yuqi.fan@hfut.edu.cn; cielo@mail.hfut.edu.cn

X. Yuan
Department of Computer Science and Engineering, University of North Texas, Denton, TX, USA
e-mail: xiaohui.yuan@unt.edu

© Springer Nature Switzerland AG 2020   29
X. Yuan, M. Elhoseny (eds.), *Urban Intelligence and Applications*, Studies in
Distributed Intelligence, https://doi.org/10.1007/978-3-030-45099-1_3

signature and digital watermarking. Blockchain achieves data tamper proof with the following methods:

1. Blockchain uses SECP256K1 signature algorithm to generate public keys for identity authentication [1, 2]. SECP256K1 is founded on the mathematical problem similar to the elliptic curve grounded on discrete logarithm solution. The mathematical problem cannot be solved theoretically and it can hardly be solved by brute-force either.
2. Blockchain is also a P2P network without a central node. All the nodes save a copy of the data block. The consistency of the block replicas can be guaranteed by consensus protocol so as to achieve the fault-tolerance and the security of the blockchain system. The consensus protocol is a vital approach to preventing data from being tampered. Popular consensus mechanisms are Proof-of-Work (PoW) [3], Proof-of-Stake (PoS) [4], Delegated PoS (DPoS) [5], Practical Byzantine Fault Tolerance (PBFT) [6], etc.
3. A block is linked to the previous block via a hash pointer so that one has to recalculate the hash value of all the successive blocks if he wants to modify a block, which increases the computational complexity.

There are some studies applying blockchain to data storage. A blockchain-based data storage framework was built using Ethereum [7]. A decentralized data storage model suitable for storing metadata was proposed to improve the security of cloud storage [8]. A blockchain-based encrypted storage network was designed, where the original data are encrypted, signed, and stored in the P2P file system, and the user can verify the integrity of the data [9, 10]. A blockchain was constructed to guarantee the security of dynamic data storage in a consensus way by building a communication channel through a key mechanism, performing data transmission on a random path in a multi-slice manner, realizing a multi-protocol synchronous operation mechanism, and completing the end-to-end encryption and secure transmission in the communication and transmission links. A decentralized privacy data management method was proposed, which uses blockchains to control access rights and effectively protect private data [11]. A secure P2P-type storage scheme was proposed to use the secret sharing and the blockchain; user data are divided into different parts, assigned to different nodes, and the node attacking the other nodes can be identified if the nodes supervise each other [12].

A wrong block in the blockchain a node maintains may cause a data inconsistency problem [13]. Some work focus on how to detect the inconsistency [14, 15]. The majority rule can handle the problem and return the correct data. However, all the successive blocks in the node are potentially wrong, which makes the node maintaining the blockchain useless. Some blockchain systems based on Hyperledger Fabric use the gossip data dissemination protocol [16] to maintain the data consistency of the nodes by transmitting information about consistency to other nodes; however, the transmission cost of the gossip method is high.

When a blockchain is used to store data to achieve data integrity, each node in the blockchain is usually required to store data so that all the nodes need to have computing and storage capacity. That is, a fat client is needed, which is often infeasible. Centralized service points make a thin client possible at the cost of introducing centralized points. In addition, a wrong block breaks data consistency, and all the successive blocks in the blockchain are potentially wrong. If the error cannot be fixed, the node maintaining the block may be useless. Existing gossip data dissemination protocol cannot solve the wrong blocks problem efficiently.

In this paper, we propose a thin client error-correcting data tamper-proofing storage framework based on blockchain. The data, the hash value of the data, and other related information are stored as blocks in the blockchain, and the clients can search the data without storing data. The contributions of this paper are as follows:

1. We investigate the problem of the thin client error-correcting tamper-proofing data storage problem, and propose a three-layer framework to store the data in the blockchain, where the framework consists of application layer, processing layer, and infrastructure layer.
2. We propose a random sampling consensus mechanism to strike a tradeoff between system decentralization and the computation power required by the traditional blockchain consensus of the network majority. A portal in the infrastructure layer receives clients requests and forwards the requests to the servers for data storage and data query. When storing data, the portal randomly selects K servers as temporary master nodes which own the right to generate and broadcast the block copy. The other nodes receive the block copy sent by each temporary master node, compare the K block copies, and keep the copy with the number of duplicated times no less than $\lceil \frac{K}{2} \rceil$. In the data query process, K servers are also randomly selected by the portal as the temporary master nodes. After verifying the queried block, the K temporary master nodes send the block data copies to the client. The client accepts the data copy with the number of duplicated times no less than $\lceil \frac{K}{2} \rceil$.
3. We propose two data tamper-proofing mechanisms, single-block validation mechanism and periodic blockchain verification mechanism, to correct the wrong blocks and ensure the data integrity. The single-block validation mechanism has two main functions: verifying the integrity of a block in a local server, and correcting a wrong block. With the periodic verification mechanism, each server periodically verifies the correctness and integrity of all blocks in the blockchain, and invokes the single-block validation to correct the wrong block.

2   Error-Correcting Data Storage Framework

In this section, we propose a thin client error-correcting data tamper-proofing storage framework based on blockchain. The framework adopts a three-layer hierarchical structure, consisting of application layer, processing layer, and infrastructure layer. The infrastructure layer consists of a portal and N servers interconnected via peer-to-peer networks. The portal receives the data access requests and forwards the requests to the servers for processing. The processing layer consists of four modules, e.g. data upload, single-block validation, periodic blockchain verification, and data query, which realizes the functions of data storage, data search, and verification.

The data upload module is responsible for appending the data submitted by the application layer to the blockchain in each server at the infrastructure layer. The single-block validation module is responsible for verifying the information of a specific block maintained in the blockchain in each server, and correcting the data in the block if it is wrong. The periodic blockchain verification module verifies the correctness of the blockchain maintained by each server on a periodic basis. The data query module receives the data search request from the application layer, searches the corresponding block and data in the blockchains, and returns the query result.

3 Modules of the Proposed Framework

The processing layer (the core of the framework) includes four modules: the data upload module, the data query module, the single-block validation module, and the periodic blockchain verification module. The rest of this section gives details of each module.

3.1 Data Upload Module

The data upload module receives the data storage request from the application layer, creates a block for the data, and stores the block in the blockchain in each server. After receiving the data storage request, the portal randomly selects $K(K \geq 3)$ servers from N servers as the temporary master nodes for the request, and returns the addresses of the selected K servers to the client. The client uploads the data to the K servers, which receive the data, generate blocks, and broadcast the hash of the generated block copies to the remaining $N - 1$ servers in the network. Each server in the network accepts the hash value with the number of duplicated times no less than $\lceil \frac{K}{2} \rceil$, and requests the generated block from one of the servers whose generated block has the same hash value as the accepted one. After verifying the received block and the last block in the local blockchain, the server appends the block to the blockchain the server maintains.

Step 1. The data upload module sends the client's data storage request to the portal in the infrastructure layer.

Step 2. The portal receives the data storage request, randomly selects K servers from N servers in the network as the temporary master nodes, and sends the addresses of the K servers to the client.

Step 3. The client uploads the data to the K selected servers after necessary processing. According to different application requirements, the client's processing of the data may include encryption, generating digital watermarks, etc.

Step 4. A new block is generated for the uploaded data by each of the K selected servers, and the hash of the block is broadcast to the remaining $N - 1$ servers.

Step 5. Each server accepts the hash value with the number of duplicated times no less than $\lceil \frac{K}{2} \rceil$.

Step 6. Each server keeps requesting the generated block from one of the servers whose generated block has the same hash value as the accepted one, and calculating the hash value of the received block, until it is the same as the previously received hash value.

Step 7. Each server calculates the hash value of the last block in the local blockchain and compares the hash value with the previous hash value in the received block. If they are the same, Step 8 is performed; otherwise, the single-block validation module of the processing layer is executed to correct the last block on the local blockchain.

Step 8. Each server appends the received block to the local blockchain.

3.2 Data Query Module

After receiving the data search request at the application layer, the framework calls the data query module to complete the data search. Upon receiving the data search request, the portal randomly selects $K(K \geq 3)$ servers from N servers as the temporary master nodes. Each of the K servers finds the queried block in the server's local blockchain. After verifying the block, each server sends the hash of the block to the client. The client accepts the hash value with the number of duplicated times no less than $\lceil \frac{K}{2} \rceil$, and requests the block from one of the servers whose queried block has the same hash value as the accepted one. The data in the received block is accepted as the query result.

Step 1. The data query module sends the data search request to the portal.

Step 2. The portal randomly selects K servers from N servers as the temporary master nodes and returns the addresses of the K servers to the client.

Step 3. The client sends the search requests to each of the K servers.

Step 4. Each of the K servers finds the block corresponding to the search request, and calls the single-block validation module to verify the block.

Step 5. Each of the K servers sends the hash of the corresponding block after verifying the block. According to different application requirements, the processing of data may include encryption, generating digital watermarks, etc.

Step 6. The client accepts the hash value with the number of duplicated times no less than $\lceil \frac{K}{2} \rceil$ as the query result.

Step 7. The client keeps requesting the block from one of the servers whose block
has the same hash value as the accepted one, and calculating the hash value
of the received block, until the calculated hash value is the same as the
previously received one.

3.3 Single-Block Validation Module

The single-block validation module performs verification of the integrity of the
block in the server and corrects the wrong block. After receiving the call for the
single-block validation module, the server sends request to obtain the hash value of
the corresponding block copy from each of the remaining $N-1$ servers, and accepts
the hash value with the number of repetition times no less than $\lceil \frac{N}{2} \rceil$. If the accepted
hash is different from the hash of the corresponding block in the local server, the
server requests the block from one of the servers whose block has the same hash
value as the accepted one. The received block is used to update the corresponding
local block.

Step 1. Upon receiving a block validation request for a specific block, the server
finds the corresponding block in the local blockchain the server maintains.
Step 2. The server broadcasts the block validation request for the specific block to
the remaining $N-1$ servers.
Step 3. The server receiving the block validation request returns the hash of the
corresponding block in the blockchain maintained by the server.
Step 4. The server sending the validation request accepts the hash value with the
number of repetition times no less than $\lceil \frac{N}{2} \rceil$.
Step 5. The server checks whether the accepted hash is the same as the hash of the
corresponding local block. If yes, the validation is successful; otherwise,
the validation fails, and Step 6 is performed.
Step 6. The server keeps requesting the block from one of the servers whose block
has the same hash value as the accepted one, and calculating the hash
value of the received block, until the calculated hash value is the same
as the previously accepted one. The server updates the block in the local
blockchains with the received block.

3.4 Periodic Blockchain Verification Module

Each server in the infrastructure layer periodically verifies the integrity of all the
blocks in the local blockchain, starting from the first block.

Step 1. The blocks in the blockchain the server maintains are numbered in
sequence, starting from 1; denote the last block as the q-th block; initialize
$p = 1$.

Step 2. Calculate the hash value of the p-th block as $h1$, and compare $h1$ with the previous block hash value, $h2$, stored in the $(p+1)$-th block; if they are the same, the p-th block is correct; otherwise, the p-th block is wrong and Step 4 is performed.

Step 3. $p = p + 1$; if $p > q - 1$, the verification is completed; otherwise, return to Step 2.

Step 4. Invoke the single-block validation module to correct the p-th block, and return to Step 2.

4   System Analysis

4.1   Data Tamper Proof Performance

1. Data tamper proof during uploading

The data are uploaded to the K randomly selected servers. The data are not uploaded through the portal, and hence the attacker cannot acquire or modify the data by attacking the portal. The attacker cannot know the specific servers that need to be attacked in advance, since the servers are randomly selected upon each data storage request by the portal. The K servers receive the data, generate blocks, and broadcast the hash of the generated block copies to the remaining $N - 1$ servers in the network. If there is a hash value with the number of duplicated times no less than $\lceil \frac{K}{2} \rceil$, each server accepts the copy.

2. Data tamper proof during query

During the data query, $K (K \geq 3)$ servers are randomly selected as the temporary master nodes, each performing the single-block validation to check the corresponding block and correct it if necessary. In addition, the data are transmitted to the client by the K servers directly without going through the portal, and hence the attacker cannot modify the data by attacking the portal. Before transmitting the data to the client, each of the K temporary master nodes may perform necessary processing, such as encryption, generating digital signature or digital watermark, etc.

3. Data tamper proof during storage

The data tamper proof during storage is guaranteed by the characteristics of the blockchain. Each block in the blockchain is linked to the previous block via the hash. When the client uploads data, each server calculates the last block hash value and compares it with the hash value of the previous block stored in the accepted block. If the two hash values are the same, the last block in the blockchain is correct; otherwise, the server performs the single-block validation to correct the wrong block.

4. Two data error-correcting mechanisms

Two data error-correcting mechanisms, single-block validation and periodic blockchain verification, ensure the integrity of data blocks. During data uploading, each server verifies the last block and correct it, if it is wrong. During data query, the temporary master nodes perform the single-block validation to check the corresponding block and correct it, if it is wrong.

Each server conducts the periodic blockchain verification to validate all the blocks in the local blockchain. When some errors are detected, the server performs the single-block validation to correct the block. If a server node has not successfully stored data or its data has something wrong, the node will synchronize or correct the data with the function of periodic blockchain verification.

4.2 Tradeoff Between System Decentralization and the Amount of Computation and Communication

The proposed framework achieves a good tradeoff between system decentralization and amount of computation.

In the proposed framework, we introduce a portal to receive the data access requests from the clients. The portal finds the servers to accommodate the data requests. K temporary master nodes are randomly selected upon each data access request. The K temporary master nodes are responsible for storing the uploaded data in the blockchains and returning the query data by searching and validating the corresponding block. The data are transmitted from the client to the servers without being forwarded by the portal during the data storage process. The data are sent from the servers to the client without the portal being an intermediate node during the data query process. That is, the data are not handled by the portal. We thereby avoid having a node or some specific nodes which have all the rights to operate data, such as block query, block broadcasting, etc., by randomly selecting the temporary master nodes. Therefore, we increase the performance of decentralization compared to PoS, DPoS, and PBFT.

With our framework, only K nodes are randomly selected to participate in the consensus process, instead of all the N servers. In addition, the random selection of the master nodes is fast and requires low computational power in the consensus process. Only hash value of the block is transmitted in the network during the consensus process for data uploading, search, and validation. Therefore, the amount of computation and communication is reduced compared to PoW, PoS, DPoS, and PBFT.

5 Conclusion

Traditional centralized and distributed data storage systems are hard to achieve data integrity. Widely used technologies, digital signature and digital watermarking, cannot protect the data from being tampered effectively. A blockchain is resistant to modification of the data via the consensus of the network majority. When a blockchain is used to store data, all the nodes, including the clients, should have powerful computing and storage capacity. Centralized service points make a thin client possible, while the centralization points are prone to be attacked. In addition, a wrong block breaks the data consistency in the system. Existing gossip data dissemination protocol cannot solve the wrong block problem efficiently. This paper proposes a thin client error-correcting data storage framework based on blockchain. The data of the clients are stored as blocks in the system to achieve the data integrity. The data requests are handled by randomly selected blockchain nodes. We propose two error-correcting mechanisms to validate and correct the wrong data blocks. We also analyze the performance of the proposed framework with respect to data tamper proof, consensus, and thin client. The proposed framework has the characteristics of lightweight client, data tamper proof, and high resource utilization in the blockchain network.

Acknowledgments This work was partly supported by the National Natural Science Foundation of China (U1836102), the Anhui Provincial Natural Science Foundation (1608085MF142), the open project of State Key Laboratory of Complex Electromagnetic Environment Effects on Electronics and Information System (CEMEE2018Z0102B), and the National Undergraduate Training Programs for Innovation and Entrepreneurship (201710359019).

References

1. Standards for efficient cryptography sec 2: recommended elliptic curve domain parameters. [Online]. Available: http://www.secg.org/sec2-v2.pdf. Accessed 20 Aug 2018
2. M. Elhoseny, H. Elminir, A. Riad, X. Yuan, A secure data routing schema for WSN using elliptic curve cryptography and homomorphic encryption. J. King Saud Univ. Comput. Inf. Sci. **28**(3), 262–275 (2016)
3. A.D. Vries, Bitcoin's growing energy problem. Joule **2**(5), 801–805 (2018)
4. Ethereum: A Next-Generation Smart Contract and Decentralized Application Platform. [Online]. Available: https://genius.com/Ethereum-ethereum-whitepaper-annotated. Accessed 20 Aug 2018
5. Delegated Proof-of-Stake Consensus. [Online]. Available: https://bitshares.org/technology/delegated-proof-of-stake-consensus/. Accessed 20 Aug 2018
6. M. Castro, B. Liskov, Practical byzantine fault tolerance and proactive recovery. ACM Trans. Comput. Syst. **20**(4), 398–461 (2002)
7. G. Zyskind, O. Nathan, A.S. Pentland, Decentralizing privacy: using blockchain to protect personal data, in *IEEE Security & Privacy Workshops*, San Jose, CA, 21–22 May 2015
8. E. Davidson, Hive mentality or blockchain bloat? New Sci. **228**(3043), 52–52 (2015). https://scholar.google.com/scholar?cluster=8368252045942210018&hl=en&num=20&as_sdt=20000005&sciodt=0,21

9. New kid on the blockchain. New Sci. **225**(3009), 7 (2015). https://www.sciencedirect.com/science/article/abs/pii/S0262407915603219

10. G. Zhou, P. Zeng, X. Yuan, S. Chen, K.-K.R. Choo, An efficient code-based threshold ring signature scheme with a leader-participant model, security and communication networks, pp. 1–7 (2017). Article ID 1915239

11. M. Gomathisankaran, X. Yuan, P. Kamongi, Ensure privacy and security in the process of medical image analysis, in *IEEE International Conference on Granular Computing* (2013), pp. 120–125

12. M. Fukumitsu, S. Hasegawa, J. Iwazaki, M. Sakai, D. Takahashi, A proposal of a secure P2P-type storage scheme by using the secret sharing and the blockchain, in *2017 IEEE 31st International Conference on Advanced Information Networking and Applications (AINA)*, Taipei, 27–29 March 2017

13. M. Dai, S. Zhang, H. Wang, S. Jin, A low storage room requirement framework for distributed ledger in blockchain. IEEE Access **6**, 22970–22975 (2018)

14. C. Anglano, R. Gaeta, M. Grangetto, Securing coding-based cloud storage against pollution attacks. IEEE Trans. Parallel Distrib. Syst. **28**(5), 1457–1469 (2017)

15. L. Buttyan, L. Czap, I. Vajda, Detection and recovery from pollution attacks in coding-based distributed storage schemes. IEEE Trans. Depend. Sec. Comput. **8**(6), 824–838 (2011)

16. Gossip data dissemination protocol. [Online] Available: http://hyperledger-fabric.readthedocs.io/en/latest/gossip.html. Accessed 20 Aug 2018

Photonic Crystal Microstrip Antenna Array Design Using an Improved Boolean Particle Swarm Optimization

Jianxia Liu, Hui Miao, Xiaohui Yuan, and Jianfang Shi

Abstract Photonic crystals have been used in antenna arrays to suppress mutual coupling. The design of microstrip antenna based on periodic photonic crystal structure is non-trivial and requires optimization of multiple factors. In this paper, we propose a Chaotic Boolean PSO algorithm for the design of microstrip antenna array with 2D mushroom photonic crystals. In our method, two different chaos sequences are employed to diversify the initialization and particle updates, which improves the particle search coverage and accelerates the convergence. The return loss and mutual coupling are used to construct the fitness function for the proposed CB-PSO. Experiments are conducted using multi-modal functions to evaluate the robustness of the proposed method against the state-of-the-art optimization methods as well as antenna design. Our results demonstrate that the proposed CB-PSO consistently achieved the best performance among state-of-the-art methods. Compared to the second best results, the improvements in CB-PSO are at least two folds. In our experiments of optimizing photonic crystal layout, CB-PSO achieves an optimized antenna design with much-improved performance. The mutual coupling is reduced by 5 dB with respect to the antenna with a full array of photonic crystal component; that is an improvement of 29.4%. In addition, the number of photonic crystal component is reduced from 48 to 24, which shows an advantage in the manufacture of photonic crystal microstrip antenna array.

Keywords Antenna array · Optimization · Particle swarm · Chaos

J. Liu · H. Miao · J. Shi
College of Information and Computer, Taiyuan University of Technology, Taiyuan, Shanxi, China

X. Yuan (✉)
Department of Computer Science and Engineering, University of North Texas, Denton, TX, USA
e-mail: xiaohui.yuan@unt.edu

© Springer Nature Switzerland AG 2020
X. Yuan, M. Elhoseny (eds.), *Urban Intelligence and Applications*, Studies in Distributed Intelligence, https://doi.org/10.1007/978-3-030-45099-1_4

1 Introduction

Because of its compact size, microstrip antenna arrays have become a key component of the miniaturized, wearable, wireless devices that communicate with exterior devices for monitoring and documenting biometric records such as heart rate, respiration rate, etc. To simplify the design of antenna and improve its performance, various techniques have been developed. Photonic crystals (PCs) are periodically structured electromagnetic media and some ranges of the frequency of an electromagnetic wave cannot propagate through the structure [1]. They are applied in the design of antenna arrays to suppress the mutual coupling between antenna arrays. The introduction of periodic photonic crystal structure into the microstrip antenna array, that is, the metal patch around the array units forming a photonic crystal grid, effectively suppresses the propagation of surface wave and enhances the radiation efficiency. The design of microstrip antenna based on periodic photonic crystal structure is non-trivial and requires optimization of multiple factors. To facilitate antenna design, several optimization methods, such as particle swarm optimization (PSO) [2] and genetic algorithm (GA) [3], have been applied [4]. These methods enable researchers to obtain intricate structures with better functionalities that could not be achieved with conventional methods.

There have been a number of bio-inspired optimization algorithms such as GA, PSO, ant colony optimization [5], immune algorithm [6], and crow search algorithm [7]. Despite different encoding structure and updating scheme, these algorithms introduce randomization at different stages to ensure coverage of the search space. On the other hand, the employment of randomization also prolongs the time to converge to an optimal solution. Hence, with a limited number of updates, these methods could get stuck in local minima [8].

To balance the success rate of reaching the global optimal and time needed to reach convergence, hybrid approaches have been developed [9, 10]. In particular, Boolean PSO integrates Boolean algorithm in PSO to update the generation with a new iterative method that differs from the binary version of particle swarm optimizations [8, 11, 12]. Boolean PSO has demonstrated a promising algorithm for dealing with various engineering design problems such as antenna design [13], designs of photonic crystals structure [11], etc. However, Boolean algorithms face challenges in high dimensional search spaces and there is room to improve efficiency.

In this paper, an improved Boolean PSO algorithm is proposed for the design of microstrip antenna array with 2D mushroom photonic crystal. In our method, two different chaos sequences are employed to diversify the initialization and particle updates, which improves the particle search coverage and accelerates the convergence. The microstrip antenna array is simulated using HFSS. The simulation results for the return loss and mutual coupling are recorded. The optimization of the array is performed under specific requirements concerning the impedance matching condition and the mutual coupling of the array elements. The simulation result shows the applicability and efficiency of the proposed technique.

The rest of this paper is organized as follows. Section 2 reviews the related work on microstrip antenna array design using optimization methods. Section 3 presents our proposed method with a description of the improved Boolean PSO and its application to the microstrip antenna array design. Section 4 discusses our results from two aspects: the performance of the proposed optimization method and its application to the design. Section 5 concludes this paper with a summary of the work.

2 Related Work

In antenna array pattern synthesis, the objective is to find the optimal weights to factors to achieve the desired radiation pattern. In the design of such antennas, photonic crystals are integrated, which are periodic dielectric structures and are capable of manipulating and controlling the flow of light. The two-dimensional PC is relatively easy to fabricate and very useful in integrated optics. In addition, when the photonic crystal structure is added to the dielectric substrate of the microstrip patch antenna, the propagation of surface waves in the substrate can be inhibited and the coupling radiation power of the antenna is suppressed, which inherently improves the performance of microstrip antenna.

Various algorithms have been adopted for reaching an optimal design including analytical methods (such as Dolph, Chebyshev, and Taylor method), numerical method (such as Powell and Conjugate gradient), and optimization methods (such as GA and PSO). A two-stage GA with a floating mutation probability was developed to design a two-dimensional (2D) photonic crystal of a square lattice with the maximal absolute band gap [14]. Since then, GA has been heavily used to enhance the performance of microstrip patch antennas (MPAs) by optimizing the bandwidth, multi-frequency, directivity, and size [15–18]. Civicioglu applied fifteen evolutionary search algorithms to solve three different circular antenna array design problems [19].

Another popular family of optimization methods is swarm intelligence. The cooperative strategy manifests PSO the concise formulation and the ease in implementation. PSO has been employed to optimize the design of array antenna and microstrip antennas such as E-shape, H-shape, and fractal antenna, which enables the efficient design of irregular antennas [20–22]. Wang et al. adopted binary particle swarm optimization and leverage IE3D to broaden the bandwidth of the C-Band patch antenna [23]. Estrada-Wiese et al. combined stochastic optimization algorithms (Random Search, PSO, and Simulated Annealing) with a reduced space-search method to obtain broadband reflecting photonic structures [24]. Integration of Boolean PSO with the scattering matrix method has also been explored [11, 25]. The result showed that the Boolean PSO has a better optimization performance for solving this problem in comparison with GA [25]. Bhaskaran et al. compared GA, PSO, and accelerated PSO to locate the feed point of a microstrip patch antenna for wireless communication [26] and concluded that PSO based method achieved better performance compared to GA.

Despite the advancement of redesigning the optimization methods for automating microstrip patch antenna prototyping, the balance of achieving global optimal and the time used in computation remains a challenge. This becomes a severe issue when the complexity of the antenna grows with more units integrated into one chip. By introducing randomness of different characteristics, this paper proposes a method that greatly improves the efficiency without sacrificing the chance of reaching a globally optimal solution.

3   Chaotic Boolean Particle Swarm Optimization

3.1   Boolean PSO Algorithm

Particle swarm optimization (PSO) is a bio-inspired algorithm that imitates birds foraging. In Boolean PSO, all variables are binary, and logical operators are used to update within generations. Velocity and position of particles are defined based on the difference between corresponding bits of two binary strings and updated using the Boolean logical operators such as "AND," "OR," and "XOR" according to the following Boolean expressions [11, 13, 25]:

$$v_{d+1} = w \bullet v_d + c_1 \bullet [p_{best,d} \oplus x_d] + c_2 \bullet [g_{best,d} \oplus x_d] \tag{1}$$

$$x_{d+1} = x_d \oplus v_d \tag{2}$$

where $\bullet$ denotes logical "AND," $+$ denotes logical "OR," and $\oplus$ denotes logical "XOR."

The velocity computed with Eq. (1) is used to determine the new d-th bit of the particle by an XOR operator. The distance between two bits represents their difference and, consequently, the velocity bit represents the change in the next step. The second and third terms of Eq. (1) calculate this distance between x_d and $p_{best,d}$, x_d and $g_{best,d}$, respectively. The connection between all terms is established with "OR" operator. The acceleration coefficients c_1 and c_2 and the inertia coefficient w are randomly decided from the system parameters p_1, p_2, and p_w, which are real numbers in the range of [0, 1]. The probabilities for being one for c_1, c_2, w are p_1, p_2, and p_w, respectively. So the effects of previous velocity v_d^* and distance g_d^* depend on these three parameters. p_w determines the dependence of the present velocity on the previous one. Acceleration parameters, p_1 and p_2, determine the self-tendency or group-tendency of the particles, respectively.

In the process of the algorithm, velocity updating follows the reverse selection algorithm of artificial immune mechanism. The maximum velocity is specified to decide the number of allowed "1s" in the calculated velocity array. To prevent the particles from moving faster than this value, each calculated velocity array is examined for the number of "1." If this number exceeds the maximum, a randomly selected "1" is set to zero, which is repeated until the maximum allowed number of "1s" is satisfied.

3.2   Chaotic Boolean PSO

To enhance the local search capabilities and balance the possible delay to the convergence, chaos sequences are adopted to increase the diversity of the initial particles and randomness in the optimization process to strengthen the capability of local research and circumvent the premature termination.

Chaos is the characteristic of non-linear systems. Incorporating chaos sequences into Boolean PSO to construct a chaotic Boolean PSO increases the diversity of the population. In this paper, Kent chaotic sequence and Henon chaotic sequence are adapted to the population initialization and perturbation of the Boolean PSO. Following the two chaotic sequences, the particle distribution is mostly even within the scope of optimization at the initialization stage, and conditionally randomly distributed at perturbation during the optimization process. The Kent chaotic sequence is computed as follows:

$$x(n + 1) = 0.9 - 1.9|x(n)| \tag{3}$$

The Henon chaotic sequence is a segmented model of the chaotic dynamic system. It is formulated as follows:

$$x(n + 1) = 1 + y(n) - ax^2(n) \tag{4}$$

$$y(n + 1) = 0.3x(n)$$

Figure 1a shows the Kent distribution with x_0 being 0.21. It is clear that the Kent chaotic map depicts a good uniformity and ergodicity within the range. The particles are mostly evenly distributed between -1 and 1. Figure 1b shows the distribution for Henon chaotic sequence with a being 1.4. The distribution describes the particle distribution ranging from -1.5 to 1.5 and the particle density is greater in the range of -0.5 and 0.5.

In the process of updating particles, Henon chaotic sequence is added to a randomly selected subset of the particles with good fitness to generate new perturbed populations of particles. As the probability density of the Henon chaotic sequence is uneven, small perturbation, i.e., within a change of 0.5, is more likely. This allows the generation of close siblings to the more promising particles. Hence, it boosts the chance of getting out of the local optima. We present the detailed steps of our proposed Chaotic Boolean Particle Swarm Optimization (CB-PSO) method in Algorithm 1. Note that the fitness function is not specified in this algorithm, which is

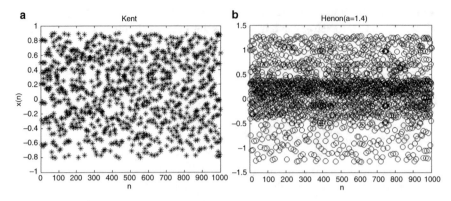

Fig. 1 The distribution of chaotic sequences of (**a**) Kent sequence $x_0 = 0.21$ and (**b**) Henon sequence $a = 1.4$

Algorithm 1 Chaotic Boolean particle swarm optimization algorithm

---

Input: Population size, subset size, and maximum iteration number
Output: The best particles g^*

Initialize particle velocity and position following Kent chaotic sequence using Eq. (3)
while less than the maximum iteration number **do**
    Update particles' velocity using Eq. (1) and position using Eq. (2)
    Compute the fitness of each particle
    Randomly select a set of particles proportional to the normalized fitness
    Perturb the position and velocity of the selected particles following Henon chaotic sequence
    using Eq. (5)
    Compute the fitness and assign the best particles to g^*
end while

---

designed according to the optimization problem. In this paper, we discuss the fitness function in the next section.

3.3   Photonic Crystal Microstrip Antenna Array

Figure 2 illustrates a diagram of an antenna array with photonic crystals in between two antenna units. The photonic crystals form a lattice of 12 by 4 array that suppresses the coupling effects. Without loss of generality, we adopt a two-port network to model the scattering parameters of the photonic crystal antenna array. The scattering matrix is defined in terms of incident and reflected waves as follows:

$$B = \begin{bmatrix} s_{11} & s_{12} \\ s_{21} & s_{22} \end{bmatrix} A$$

Fig. 2 Diagram of a two-unit antenna array with photonic crystal structure

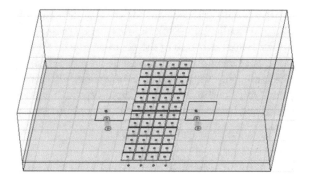

where $B = [b_1 \ b_2]^T$ and $A = [a_1 \ a_2]^T$ are vectors of reflected waves and incident waves, respectively. In the scattering matrix, s_{11} is the input voltage reflection coefficient, s_{21} is the forward voltage gain, s_{12} is the reverse voltage gain, and s_{22} is the output voltage reflection coefficient. Hence, the antenna design is to optimize the spatial arrangement of the components. The goal is to arrange the photonic crystals such that coupling through surface wave between two antennas is suppressed and radiation power is maximized.

There are a number of factors that affects the antenna performance such as input impedance, radiation patterns, mutual coupling, and return loss, etc. Among these factors, mutual coupling and return loss are the crucial ones. The return loss is measured by the reflection coefficient, i.e., s_{11}, which is the ratio between the amplitude of reflected wave and incident wave. Return loss indicates the character of impedance match and the operating bandwidth is the frequency range with over 90% incident power delivered to the antenna. The mutual coupling is measured by the transmission coefficient, i.e., s_{21}, which is the ratio between the amplitude of transmitted wave and incident wave at the feeding point. The transmitted energy can be thought of as the coupling energy between the units of the antenna array. Specifically, it is expected that the return loss of a functioning antenna must be less than the cutoff loss (i.e., -10 dB) and the mutual coupling needs to be minimized.

The fitness function of CB-PSO is the product of mutual coupling and rectified return loss as follows:

$$f(\Gamma) = s_{21}(\Gamma; w)e^{R(\Gamma; w, \hat{s})} \tag{5}$$

where Γ denotes the spatial structure of photonic crystal array, w denotes the frequency, and $R(\cdot)$ is the rectifier function in the following form:

$$R(\Gamma; w, \hat{s}) = \max(0, s_{11}(\Gamma; w) - \hat{s})$$

$\hat{s}$ is the cutoff return loss. For an acceptable return loss, -10 dB is usually used for $\hat{s}$. That is, an antenna with the frequency response of the return loss that is below

-10 dB has no influence on the optimization objective. To allow a greater emphasis on s_{11}, a smaller cutoff return loss can be specified.

The spatial structure of the photonic crystal array is encoded with binary strings. If we treat each photonic crystal as an independent element, for a twelve by four array, six bits are sufficient to represent all combinations. In our experiments, primitive cells that include multiple crystals are used, which reduced the number of bits and hence the complexity of the optimization problem.

4   Experimental Results and Discussion

All methods used in our experiments are implemented with MATLAB. We evaluate our method in two aspects: performance of convergence to global optima using numerical functions and performance of antenna using the antenna simulator ANSYS High-Frequency Structure Simulator (HFSS) (http://www.ansys.com).

4.1   Numerical Evaluation for CB-PSO

Four testing functions (Rosenbrock, Alpine function, Shaffer's F6, and Camel function) are used to evaluate our proposed CB-PSO and a comparison study is conducted using Boolean PSO and binary GA. These functions fluctuate and have many local optima. The formula of these functions are as following:

1. Rosenbrock: $f(x) = 100(x_1^2 - x_2)^2 + (1 - x_1)^2$, where $x_1, x_2 \in [-2.048, 2.048]$
2. Alpine Function: $f(x) = \sum_i |x_i \sin(x_i) + 0.1x_i|$, where $x_i \in [-100, 100]$
3. Shaffer's F6: $f(x) = 0.5 - \frac{\sin^2 \sqrt{(x_1^2 + x_2^2)} - 0.5}{(1 + 10^{-3}(x_1^2 + x_2^2))^2}$, where $x_i \in [-100, 100]$
4. Camel: $f(x) = (4 - 2.1x_1^2 + x_1^4/3)x_1^2 + x_1x_2 + 4(x_2^2 - 1)x_2^2$, where $x_i \in [-100, 100]$

All methods are evaluated for 20 repetitions with random initialization. CB-PSO and Boolean PSO have the same initial population of 80 particles, and maximum allowed number of "1s" is 5, p_1 and p_2 are 0.5, and p_w is 0.1. The maximum number of iterations for CB-PSO and BPSO is 100. The maximum generation of GA is 100 generations, which is equivalent to the maximum number of iterations for PSO methods. In the evaluation of GA, 80 chromosomes in each generation are used and each chromosome contains 10 elements. The crossover probability in GA is 0.9 and the mutation probability is 0.2.

Table 1 reports the average optimization results of the three methods and the distances to the true optima. The best results are highlighted in boldface font. In the tests with all functions, our proposed CB-PSO consistently achieved the minimum distance to the true optima. In two cases (Shaffer's F6 and Camel), CB-

Table 1 Optimization results and distance to the true optimal

Testing function	Method	Global optimal	Resulted optimal	Absolute distance
Rosenbrock	CB-PSO	0	6.50E−06	**6.50E−06**
	BPSO		1.97E−04	1.97E−04
	GA		2.56E−03	2.56E−03
Alpine	CB-PSO	0	1.96E−07	**1.96E−07**
	BPSO		1.35E−05	1.35E−05
	GA		8.58E−04	8.58E−04
Shaffer's F6	CB-PSO	1	1	**0**
	BPSO		9.96E−01	4.20E−03
	GA		1.0049	4.90E−03
Camel	CB-PSO	−1.031628	−1.031628	**0**
	BPSO		−1.02878	2.85E−03
	GA		−1.02481	6.82E−03

PSO successfully reached the global optima within 100 iterations. Compared to the second best results (underlined in the table), the improvements of CB-PSO are at least two folds.

Figure 3 depicts the typical fitness curve with respect to generations/iterations of the three methods. Overall, GA converges within a smaller number of generations. However, GA results in a local optimal at the convergence. In the case of Alpine function, GA ends up with a much less optimal solution, which is mostly due to the relative large mutation rate. It is interesting to note that although randomness is introduced into the CB-PSO method using two chaos sequences, the converge speed of CB-PSO is highly competitive in comparison to PSO. For all cases, CB-PSO converges within 50 iterations.

4.2 Photonic Crystal Optimization for Microstrip Antenna Array

In our experiments of photonic crystal optimization, the antenna array is placed on a substrate of relative permittivity four, i.e., $\epsilon_r = 4$. The size of each patch of metal foil (antenna) is 9.23 mm×9.23 mm; the size of the mushroom patch is 4.07 mm×4.07 mm, the periodic interval is 0.37 mm; the excitation source is in coaxial feed mode, and the length from coaxial center to the left side of the patch is 2.93 mm. The conductivity of the metallic conductor is 5.7 E+7 S/m, and the antenna is designed to match the resistance of 50Ω.

Figure 4 shows a two-unit antenna array with and without photonic crystal as well as the frequency response of return loss and mutual coupling of the antennas. The horizontal solid line marks the −10 dB, i.e., the maximum allowed return loss for an operational antenna, which decides a frequency range from the curve of s_{11}.

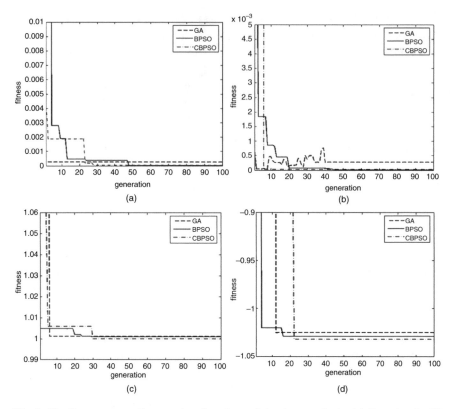

Fig. 3 The fitness across all generations/iterations of the three methods. (**a**) Rosenbrock. (**b**) Alpine function. (**c**) Shaffer's F6. (**d**) Camel

Within this frequency range (marked with vertical dash lines in Fig. 4b and d), the mutual coupling for the antenna with and without photonic crystal is in the range of -28.75 dB to -34.75 dB and -13.5 dB to -14.5 dB, respectively. It is clear that by adding photonic crystal, the performance of the antenna is significantly improved; the reduction of mutual coupling is 17 dB based on the median value.

To optimize the placement of photonic crystals, we employ the proposed CB-PSO method and the primitive cell is used to simplify the spatial arrangements. Each primitive cell consists of a number of consecutive photonic crystals within an N by N window. Hence, the optimization of a photonic crystal structure in an antenna becomes the arrangement of primitive cells. Note that there is no overlap between primitive cells. An example is shown in Fig. 5, where 16 metallic patches in a four by four window form a primitive cell. This mushroom photonic crystal structure includes three primitive cells.

Figure 6 shows the optimized photonic crystal structure using primitive cells of 4 by 4 and its frequency response. The optimal arrangement of photonic crystal cells is achieved with the particle [1, 1, 0]. The structure of the antenna is depicted

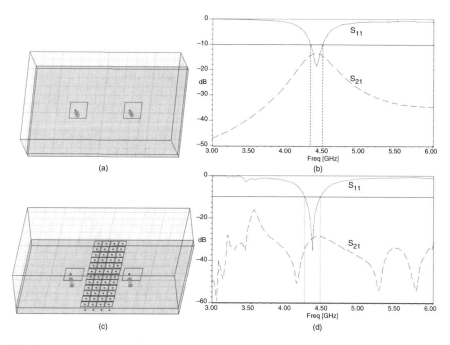

(a)   (b)

(c)   (d)

Fig. 4 The frequency response of return loss s_{11} and mutual coupling s_{21} of a two-unit photonic crystal antenna array. (**a**) and (**c**) show the layout of the metallic patches. (**b**) and (**d**) depict the frequency response of the antenna array in (**a**) and (**c**), respectively

Fig. 5 A primitive cell with 16 metallic patches

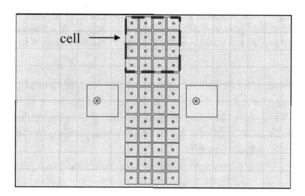

in Fig. 6a and the frequency responses of the return loss (s_{11}) and mutual coupling (s_{21}) are shown in Fig. 6b. Within the frequency range that satisfies $s_{11} < -10\,\mathrm{dB}$, the mutual coupling s_{21} is in the range of $-16\,\mathrm{dB}$ to $-24\,\mathrm{dB}$. In contrast to the two-unit antenna array without photonic crystal, the reduction is 6 dB. In addition, the antenna design resulted from our method requires only two-thirds of the photonic crystals.

When four photonic crystal components in a two by two window are used as a primitive cell, the photonic crystal structure includes 12 primitive cells. Figure 7a

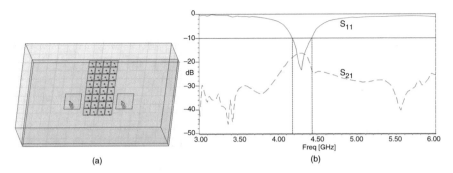

(a)   (b)

Fig. 6 The structure and operation results of an antenna array with three primitive cells. (a) Optimized geometry of the antenna array with three primitive cells. (b) The frequency response of return loss and mutual coupling of this antenna

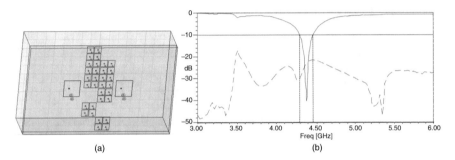

(a)   (b)

Fig. 7 The structure and operation results of an antenna array with 12 primitive cells. (**a**) Optimized geometry of the antenna array with 12 primitive cells. (**b**) The frequency response of return loss and mutual coupling of the optimized antenna with 12 primitive cells

depicts the optimized antenna layout. The number of photonic crystal components is 32, that is, two-thirds of the full array of photonic crystal used in the antenna shown in Fig. 4c. Following the same return loss criterion, the range of mutual coupling is -22 dB to -30 dB, which is very competitive to the antenna using a full array of the photonic crystal.

Using a single photonic crystal component as a primitive cell, we further evaluate the optimization of the antenna layout. The resulted photonic crystal layout is shown in Fig. 8a and its frequency response is shown in Fig. 8b. The range of the return loss of this optimized antenna is -30 dB to -42 dB, which is better than using the full array of the photonic crystal. The number of photonic crystal components is 24, which reduces the number of photonic crystal by half compared to the antenna shown in Fig. 4c.

Table 2 summarizes the performance factors of the optimized antennas. In this table, the two-unit antenna array without mushroom photonic crystal is used as the baseline. With 48 photonic crystal components, the medium mutual coupling is reduced by 17 dB. Using primitive cells in our CB-PSO method, a less number

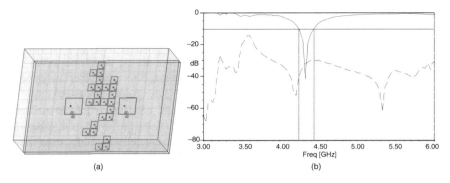

(a)   (b)

Fig. 8 The optimized antenna array of the photonic crystal structure. (**a**) The rearrangement of primitive cells. (**b**) The frequency response of the return loss and mutual coupling of the optimized antenna

Table 2 Return loss and mutual coupling of the photonic crystal microstrip antenna arrays

Antenna	Mutual coupling (dB)			Reduction (dB)	# of photonic crystal
	Min	Max	Medium		
Two-unit antenna	−14.5	−13.5	−14	–	–
w. full array of photonic crystal	−34.75	−28.75	−31	17	48
w. optimized 4 by 4 primitive cell	−24	−16	−20	6	32
w. optimized 2 by 2 primitive cell	−30	−22	−26	12	32
w. optimized 1 by 1 primitive cell	−42	−30	−36	22	24

The dashes denote no corresponding values

of photonic crystal components is used and competitive performance is achieved. An advantage of using primitive cells is the simplification of the particle encoding because a shorter binary string is usually used. When the primitive cell contains a single photonic crystal component, we can achieve a much significant reduction of the mutual coupling, which is 22 dB. This is superior to the case of using a full array of the photonic crystal. In addition, the number of photonic crystal component is reduced to 24, i.e., half of a full array. It is evident that the optimized mushroom photonic crystal layout with a much less number of photonic crystal components achieved with our proposed CB-PSO method effectively suppresses the mutual coupling between antennas and improves the overall performance. The improvement is at 29.4% in terms of the medium mutual coupling loss compared to the antenna with a full array of photonic crystal components.

5   Conclusion

The design of microstrip antenna based on periodic photonic crystal structure is non-trivial and requires optimization of multiple factors. To balance the success rate of reaching the global optimal and time needed to reach convergence, we propose a Chaotic Boolean PSO algorithm for the design of microstrip antenna array with 2D mushroom photonic crystals. In our method, two different chaos sequences are employed to diversify the initialization and particle updates, which improves the particle search coverage and accelerates the convergence. The return loss and mutual coupling are used to construct the fitness function for the proposed CB-PSO.

Experiments are conducted using widely used multi-modal functions to evaluate the robustness of the proposed method against the state-of-the-arts optimization methods. The proposed CB-PSO consistently achieved the minimum distance to the true optima. In two cases (Shaffer's F6 and Camel), CB-PSO successfully reached the global optima within 100 iterations. Compared to the second best results, the improvements in CB-PSO are at least two folds. In addition, the converging speed of CB-PSO is highly competitive in comparison to BPSO. For all cases, CB-PSO converges within 50 iterations.

In our experiments of antenna design, the mutual coupling of antenna arrays with optimized photonic crystal component layout is compared with that of the antenna with a full array of photonic crystal components. It is demonstrated that as we reduce the size of the primitive cells, the optimized antenna achieves a better performance. When a primitive cell consists of a single photonic crystal component, the optimized antenna array exhibits a much-improved performance. The mutual coupling is reduced by 5 dB with respect to the antenna with a full array of photonic crystal component; that is an improvement of 29.4%. In addition, the number of photonic crystal component is reduced from 48 to 24, which shows an advantage in the manufacture of photonic crystal microstrip antenna array.

References

1. E. Yablonovitch, Inhibited spontaneous emission in solid-state physics and electronics. Phys. Rev. Lett. **58**, 2059–2062 (1987)
2. J. Kennedy, W.M. Spears, Matching algorithms to problems: an experimental test of the particle swarm and some genetic algorithms on the multimodal problem generator, in *IEEE International Conference on Evolutionary Computation Proceedings*, May 1998, pp. 78–83
3. J. McCall, Genetic algorithms for modelling and optimisation. J. Comput. Appl. Math. **184**(1), 205–222 (2005)
4. A.S.S. Ramna, Design of rectangular microstrip patch antenna using particle swarm optimization. Int. J. Adv. Res. Comput. Commun. Eng. **2**, 2918–2920 (2013)
5. M. Dorigo, C. Blum, Ant colony optimization theory: a survey. Theoret. Comput. Sci. **344**(2), 243–278 (2005)
6. S. Al-Sharhan, Artificial immune systems - models, algorithms and applications. Int. J. Res. Rev. Appl. Sci. **3**, 118–131 (2010)
7. A. Askarzadeh, A novel metaheuristic method for solving constrained engineering optimization problems: Crow search algorithm. Comput. Struct. **169**, 1–12 (2016)
8. D. Gorse, Binary particle swarm optimisation with improved scaling behaviour, in *European Symposium on Artificial Neural Networks, Computational Intelligence and Machine Learning*, April 2013, pp. 24–26

9. S. Prakash, D.P. Vidyarthi, A hybrid immune genetic algorithm for scheduling in computational grid. Int. J. Bio-Inspired Comput. **6**(6), 397–408 (2014)
10. M.S. Kıran, E. Özceylan, M. Gündüz, T. Paksoy, A novel hybrid approach based on particle swarm optimization and ant colony algorithm to forecast energy demand of Turkey. Energy Convers. Manage. **53**(1), 75–83 (2012)
11. A. Marandi, F. Afshinmanesh, P.P.M. So, Design of a highly focused photonic crystal lens using Boolean particle swarm optimization, in *IEEE Lasers and Electro-Optics Society Annual Meeting Conference Proceedings*, October 2007, pp. 931–932
12. S. Gunasundari, S. Janakiraman, S. Meenambal, Velocity bounded Boolean particle swarm optimization for improved feature selection in liver and kidney disease diagnosis. Exp. Syst. Appl. **56**, 28–47 (2016)
13. F. Afshinmanesh, A. Marandi, M. Shahabadi, Design of a single-feed dual-band dual-polarized printed microstrip antenna using a Boolean particle swarm optimization. IEEE Trans. Antenn. Propag. **56**(7), 1845–1852 (2008)
14. L. Shen, Z. Ye, S. He, Design of two-dimensional photonic crystals with large absolute band gaps using a genetic algorithm. Phys. Rev. B **68**, 1–5 (2003)
15. Paras, R.P.S. Gangwar, Design of compact and multiband antenna array using genetic algorithm optimization. **6**, 221–227 (2011)
16. J.W. Jayasinghe, D.N. Uduwawala, A novel miniature multi-frequency broadband patch antenna for WLAN applications, in *IEEE 8th International Conference on Industrial and Information Systems*, December 2013, pp. 361–363
17. J.M.J.W. Jayasinghe, J. Anguera, D. Uduwawala, Genetic algorithm optimization of a high-directivity microstrip patch antenna having a rectangular profile. Radioengineering **22**, 700–707 (2013)
18. M. Lamsalli, A. El Hamichi, M. Boussouis, N. Touhami, T. Elhamadi, Genetic algorithm optimization for microstrip patch antenna miniaturization. Progr. Electromagn. Res. Lett. **60**, 113–120 (2016)
19. P. Civicioglu, Circular antenna array design by using evolutionary search algorithms. Progr. Electromagn. Res. B **54**, 265–284 (2013)
20. J. Nanbo, R.-S. Yahya, Particle swarm optimization for antenna designs in engineering electromagnetics. J. Artif. Evol. Appl. **2008**, 10 pp. (2008)
21. S. Rani, A.P. Singh, On the design and optimisation of new fractal antenna using PSO. Int. J. Electron. **100**(10), 1383–1397 (2013)
22. C. Rai, Optimization of h-shape micro strip patch antenna using PSO and curve fitting. Int. J. Res. Appl. Sci. Eng. Technol. 993–996 (2017). https://doi.org/10.22214/ijraset.2017.10142
23. N.Z. Wang, X.B. Wang, J.D. Xu, Design of a novel compact broadband patch antenna using binary PSO. Microw. Opt. Technol. Lett. **54**(2), 434–438 (2012)
24. D. Estrada-Wiese, E.A. del Rio-Chanona, J.A. del Rio, Stochastic optimization of broadband reflecting photonic structures. Sci. Rep. **8**(1), 1193 (2018)
25. F. Afshinmanesh, Design of photonic crystals and binary supergratings using Boolean particle swarm optimization. PhD thesis, University of Tehran, 2006
26. S. Bhaskaran, R. Varma, J. Ghosh, A comparative study of GA, PSO and APSO: feed point optimization of a patch antenna. Int. J. Sci. Res. Publ. **3**, 1–5 (2013)

Remote Sensing Image Fusion Using Improved IHS and Non-subsampled Contourlet Transform

Yong Zeng, Wei Yi, Jianan Deng, Weirong Chen, Shenghao Xu, and Shusong Huang

Abstract Low-altitude security monitoring and spatial visualization can make use of high spatial resolution remote sensing images, and pixel-level image fusion techniques can be used to combine panchromatic and multispectral images to generate composite images with high resolution and rich spectral information. The IHS transform fusion is one of the most widely used techniques for image fusion. However, the IHS transform fusion brings spectral distortion. In order to develop new image fusion methods, it is necessary to investigate the spectral features of the original images from different sensors. In this study, high-resolution panchromatic images were reconstructed to improve IHS transform based on GF-2 satellite images. The NSCT transform was used in order to separate details and spectral information. A synthetic index (SI) for assessing fidelity was proposed with consideration of average gradient, entropy, correlation coefficient, and spectral distortion. Results show that, in urban areas, the SI of improved IHS method increases from 2.75 to 4.30, and the SI of the hybrid method (improved IHS + NSCT method) increases from 6.68 to 6.93. In addition, the proposed method helps to improve the SI from 1.10 to 3.80 and the NSCT from 6.00 to 7.46 for vegetation-covered areas. Thus, the improved IHS transform would maintain spectral fidelity and significantly improve the vegetation spectral information.

Keywords Image fusion · Contourlet · Low-altitude security · Spectral features

Funded by Qian Xuesen Youth Innovation Fund Project of China Aerospace Science and Technology Group Corporation (No.17 Zhengzi [2017]).

Y. Zeng · W. Yi (✉) · J. Deng · W. Chen · S. Xu · S. Huang
China Center For Resources Satellite Data and Application, Beijing, China

© Springer Nature Switzerland AG 2020
X. Yuan, M. Elhoseny (eds.), *Urban Intelligence and Applications*, Studies in Distributed Intelligence, https://doi.org/10.1007/978-3-030-45099-1_5

1   Introduction

Remote sensing satellites usually carry panchromatic and multispectral cameras to acquire surface information. In terms of panchromatic images, they have high spatial resolution and rich details, but unitary spectral information; for multispectral images, which contain a lot of spectral information, however, their spatial resolution is lower than panchromatic images. Low-altitude security remote sensing monitoring and spatial visualization require high spatial resolution and rich spectral images. Pixel-level image fusion techniques can combine panchromatic and multispectral images to generate composite images with high spatial resolution and rich spectral information [1–3].

There are many mature algorithms for data fusion. Traditional fusion algorithms include Brovey transform, IHS transform, PCA transform, and other methods based on multiplication or spatial operation [4, 5]. IHS transform can only fuse three bands of images, and bring spectral distortion while improving spatial resolution. Many scholars have proposed improved algorithms on account of IHS transform. The FIHS algorithm presented by Tu replaces IHS multiplication operation with addition, which speeds up the fusion but does not improve the spectral distortion [6].Afterward, Tu and Choi have improved the I component and reduced spectral distortion by introducing the near-infrared band and achieved a balance between spatial details and spectral information by adjusting parameters [7, 8]. In recent years, more research has focused on multi-scale transform analysis to improve I component by using Wavelet transform, Curvelet transforms, NSCT transform, and other methods to preserve spectral information [9–11].

In order to preserve spectral information, the previous studies mainly focused on I component in IHS transform and neglected the panchromatic image by replacing the I component. Although I component and panchromatic images are basically interchangeable, their spatial resolution and physical meaning are completely different. Adjusting I component by using parameters has poor robustness which depends entirely on the statistical characteristics of data, and it is poorly interpretable, thus it has no definite meaning [12]. In this paper, a new improved method of IHS transform is proposed, which is based on the spectral characteristics of multispectral and panchromatic images to fuse images by simulating high-resolution panchromatic image. This method is different from the previous method of determining I component by adjusting parameters, and its physical implication is more concise. The spectral distortion of IHS transform is caused by the incomplete overlap of spectral range between panchromatic image and multispectral image. Therefore, it is theoretically possible to improve the spectral information of fusion results by simulating a panchromatic image with three spectral ranges of red, green, and blue. In addition, NSCT transform can decompose the image into low pass sub-band and directional bandpass sub-band, and effectively distinguish the detail information from spectral information [13, 14]. Using this transform to inject the detail information of simulated panchromatic image into I component, the spectral distortion of IHS transform can be further reduced.

2   Remote Sensing Image Fusion Procedure

2.1   Traditional IHS Transform Image Fusion

IHS transform image fusion transforms multispectral images from RGB space to IHS space, while R (red), G (green), and B (blue) components are converted to I (brightness), H (hue), and S (saturation) components, in which I component stores spatial details of the image [17–21]. Replacing I component with panchromatic image and then returning to RGB space by IHS inverse transform with H and S component can improve the spatial resolution of a multispectral image. The formulas to transform RGB space into IHS space are as follows:

$$\begin{bmatrix} I \\ V_1 \\ V_2 \end{bmatrix} = \begin{bmatrix} \frac{1}{3} & \frac{1}{3} & \frac{1}{3} \\ -\frac{\sqrt{2}}{6} & -\frac{\sqrt{2}}{6} & \frac{2\sqrt{2}}{6} \\ \frac{1}{\sqrt{2}} & -\frac{1}{\sqrt{2}} & 0 \end{bmatrix} \begin{bmatrix} R \\ G \\ B \end{bmatrix} \tag{1}$$

$$H = \tan^{-1}\left(\frac{V_2}{V_1}\right) \tag{2}$$

$$S = \sqrt{V_1{}^2 + V_2{}^2} \tag{3}$$

In the formula, R, G, and B are red, green, and blue multispectral images, respectively, and V_1 and V_2 are intermediate variables.

The formulas for calculating the transformation from IHS space to RGB space after replacing I component with a panchromatic image are as follows:

$$\begin{bmatrix} R_{\text{new}} \\ G_{\text{new}} \\ B_{\text{new}} \end{bmatrix} = \begin{bmatrix} 1 & -\frac{1}{\sqrt{2}} & \frac{1}{\sqrt{2}} \\ 1 & -\frac{1}{\sqrt{2}} & -\frac{1}{\sqrt{2}} \\ 1 & \sqrt{2} & 0 \end{bmatrix} \begin{bmatrix} \text{PAN} \\ V_1 \\ V_2 \end{bmatrix} \tag{4}$$

In the formula, R_{new}, G_{new}, and B_{new} are red, green, and blue multispectral images fused by IHS transform, respectively, and PAN is the panchromatic image.

2.2   Improved IHS Transform Image Fusion

Traditional IHS transform uses red, green, and blue bands for conversion. Fast IHS transform converts multiplication operation into addition operation. The operation time is improved, but only the calculation formula is deformed and the result is unchanged. The deformed formula is as follows:

$$
\begin{bmatrix} R_{\text{new}} \\ G_{\text{new}} \\ B_{\text{new}} \end{bmatrix} = \begin{bmatrix} 1 & -\frac{1}{\sqrt{2}} & \frac{1}{\sqrt{2}} \\ 1 & -\frac{1}{\sqrt{2}} & -\frac{1}{\sqrt{2}} \\ 1 & \sqrt{2} & 0 \end{bmatrix} \begin{bmatrix} I + (\text{PAN} - I) \\ V_1 \\ V_2 \end{bmatrix} = \begin{bmatrix} R + (\text{PAN} - I) \\ G + (\text{PAN} - I) \\ B + (\text{PAN} - I) \end{bmatrix} \quad (5)
$$

More improved algorithms deform the I component to improve spectral distortion. For example, in reference paper [7], the near-infrared band is introduced to improve I component:

$$
I = \frac{R + G + B + \text{NIR}}{4} \quad (6)
$$

In reference paper [8], considering the difference between G and B band, the best weight value is obtained from 92 sets of data using IKONOS image, and I component after distortion is as follows:

$$
I = \frac{R + 0.75 \times G + 0.25 \times B + \text{NIR}}{4} \quad (7)
$$

In reference paper [15], it reconstructs I component by the spectral characteristics of the sensor, the way of reconstruction is:

$$
I = \frac{\sum_{i=1}^{n} X_i \times \left(b_{\text{up}}^i - b_{\text{down}}^i \right)}{\sum_{i=1}^{n} \left(b_{\text{up}}^i - b_{\text{down}}^i \right)} \quad (8)
$$

In the formula, i is the serial number of bands, n is the total number of bands, X_i is the image of the i band, b_{up}^i is the upper limit of the spectral range of the i band, and b_{down}^i is the lower limit of the spectral range of the i band.

The commonly used improved IHS algorithms are listed above. From formula (5), the detail information added by IHS transform fusion image comes from the difference between the PAN image and the I component. Formula (6) and formula (7) redefine the I component by simply adding NIR component. Formula (8) reconstructs I component by spectral characteristics of sensor. The improved IHS method proposed in this study also utilizes the spectral characteristics of the sensor, but unlike the reconstruction of the low resolution I component, this study reconstructs the simulated high-resolution panchromatic image.

IHS transform was first successfully applied to multispectral satellite images of three bands (SPOT-1 satellite images), but it did not work well for satellite images of four bands (IKONOS images). The reason is that the spectral range of the panchromatic image of the SPOT-1 satellite image is highly consistent with that of the multispectral image, and any combination of three bands of IKONOS image cannot be consistent with that of panchromatic image. The GF-2 satellite images used in this study are similar to the IKONOS band settings; if

IHS transform is successfully applied to GF-2 satellite images, it is necessary to simulate panchromatic images which are basically identical to the spectral ranges of three bands.

The spectral range of the panchromatic image of the GF-2 satellite is 0.45–0.90 μm. For multispectral bands, blue band image is 0.45–0.52 μm, green band image is 0.52–0.59 μm, red band image is 0.63–0.69 μm, and near-infrared band image is 0.77–0.89 μm. According to the band setting, the spectral range of panchromatic images is highly consistent with the multispectral image. However, if the IHS fusion results reflect the true color of RGB space, the overlapping part of spectrum range of panchromatic image with the near-infrared band of multispectral band is redundant. Panchromatic image is spectral reflectance intensity of four bands, and multispectral image is the spectral reflectance intensity of a single band. This study simulates a new panchromatic image by using band weighting method. The specific calculation formula is as follows:

$$w_i = \frac{\left(b_{up}^i - b_{down}^i\right)}{\sum\limits_{i=1}^{4}\left(b_{up}^i - b_{down}^i\right)} \tag{9}$$

$$f_{actor} = \frac{w_B \times B + w_G \times G + w_R \times R}{w_B \times B + w_G \times G + w_R \times R + w_{NIR} \times \text{NIR}} \tag{10}$$

$$\text{PAN}' = f_{actor} \times \text{PAN} \tag{11}$$

w_i is the weight of band, f_{actor} is the proportion of reflective intensity of simulated red, green, and blue bands in the panchromatic image, and PAN$'$ represents a simulated panchromatic image.

f_{actor} is a number less than 1, the dynamic range of the simulated panchromatic image histogram becomes narrower. In order to keep the same amount of information as the original panchromatic image, the simulated panchromatic image is transformed by histogram transformation, so that the mean and variance of the simulated panchromatic image are as close as possible to the original panchromatic image. The calculation formula is as follows:

$$\text{PAN}_{new} = \frac{\sigma_{PAN}}{\sigma_{PAN'}} \times \left(\text{PAN}' - \mu_{PAN'}\right) + \mu_{PAN} \tag{12}$$

In the formula, PAN$_{new}$ is the final simulated panchromatic image, σ_{PAN} is the variance of the PAN image, $\sigma_{PAN'}$ is the variance of the PAN$'$ image, μ_{PAN} is the mean of the PAN image, and $\mu_{PAN'}$ is the mean of the PAN$'$ image.

In this study, PAN in formula (4) will be replaced by PAN$_{new}$. The difference between PAN$_{new}$ and PAN is that PAN$_{new}$ does not contain luminance information of the near-infrared band. Therefore, the difference between PAN$_{new}$ and I component mainly reflects detail information in the panchromatic band, which has less influence on spectral information and thus reduces spectral distortion.

2.3 NSCT Transform

The multi-scale transform analysis method has been widely used in remote sensing image fusion. Wavelet transform was first applied in remote sensing image fusion because it has the decomposition ability of multi-scale and multi-resolution, which can separate the high-frequency details and low-frequency information in the image and solve the spectral distortion problem. However, as the wavelet transform does not have multidirectional and translation invariance, it will generate pseudo-Gibbs phenomenon. In [16], NSCT transformation theory overcomes the above problems; its core idea is to use a non-subsampled Laplacian pyramid filter bank (non-subsampled pyramid filter bank, NSPFB) and the non-subsampled direction filter bank (NSDFB) for decomposing images in multi-scale and multiple directions. NSPFB decomposes image in multi-scale to ensure the multi-resolution of image transformation; NSDFB decomposes the sub-band images in different scales and directions to ensure the multidirectional characteristics of image transformation, so as to obtain the sub-band images in different scales and directions.

2.4 Image Fusion Based on NSCT and Improved IHS

NSCT transform can separate the high-frequency details from the low-frequency information of the image, and preserve the details in a multi-scale and multidirectional way. The improved IHS transform can make the spectral intensity of the simulated panchromatic image close to I component of IHS transform. We decompose the simulated panchromatic image and the I component and get the high and low frequency components. The high frequency components of the I component are replaced by the high frequency component of the simulated panchromatic image. By applying the inverse transformation, we enhance the spatial resolution of the fusion results while avoiding spectral distortions. The steps of fusion are as follows:

1. Using panchromatic high-resolution image as the base image, the multispectral image is registered spatially, and the registration accuracy is better than 1 pixel;
2. Resampling the multispectral image to achieve the spatial resolution level of panchromatic image. This step is the only physical sampling of the multispectral image, but the recognition ability of actual surface features is not improved;
3. The IHS spatial transformation in R, G, and B bands of multispectral images is used to obtain I component reflecting the brightness information of images;

4. Using panchromatic, red, green, blue, and near-infrared of multispectral images and spectral characteristics of satellites, the simulated panchromatic image is calculated by the formula (11). Formula (12) is used to make the histogram distribution of the simulated panchromatic image close to I component;

5. Using NSCT transform to separate histogram to match the simulated panchromatic image and high-frequency details and low-frequency information of I component, respectively;

6. To replace high-frequency details of I component by high-frequency details of histogram matching the simulated panchromatic image, and PAN_{new} is generated by NSCT inverse transformation;

7. To replace PAN by PAN_{new} and perform IHS inverse transform using formula (4).

3   Results and Evaluation

3.1   Evaluation Indicators

1. Average gradient A_G. It is used to describe the clarity of the image, reflecting the degree of image detail contrast and texture features. The larger the average gradient, the clearer the image is.

2. Information entropy E. The information entropy of an image is a piece of average information describing the image. It is an important indicator to measure the richness of image information. The larger the information entropy value, the larger the amount of information contained in the fusion result image.

3. Correlation coefficient C_C. Reflecting the closeness of the linear relationship between the fused image and the original image, the larger the correlation coefficient, the closer the fused image is to the source image.

4. Spectral distortion S_D. It indicates the loss of spectral information of the fused image relative to the original image, reflecting the degree of distortion of the fused image. The smaller the spectral distortion is, the more the fused image retains the spectral information of the original image, the brightness is close and the spectral distortion is small.

5. Comprehensive indicator S_I. The average gradient, information entropy, correlation coefficient, and spectral distortion reflect the change in quality after image fusion from different aspects: the average gradient and the information entropy main reflect the detailed information of the fusion results; the correlation coefficient and the spectral distortion main reflect the color fidelity of the fusion results. This study proposes a new objective evaluation index S_I: theoretically, the larger the average gradient, the information entropy, and the correlation coefficient, the smaller the spectral distortion value, and the better the fused image. The formula is as follows:

$$S_I = \frac{A_G \times E \times C_C}{S_D} \tag{13}$$

3.2   Fusion Result Evaluation and Analysis

This research uses four fusion algorithms for comparative analysis. There is an IHS algorithm, improved IHS algorithm, NSCT + IHS algorithm, and NSCT+ improved IHS algorithm. The experimental data is acquired by the GF-2 satellite. The IHS algorithm uses three bands. The near-infrared band is not used in fusion, and the biggest changes feature type in the near-infrared band is vegetation. Therefore, the comparative data in this research uses a rich informative urban area and a relatively single vegetation area.

Figure 1 shows the fusion image in a city area, Fig. 1a is panchromatic image, Fig. 1b is multispectral image, Fig. 1c is IHS fusion image, Fig. 1d is improved IHS fusion image, Fig. 1e is NSCT + IHS fusion image, Fig. 1f is NSCT+ improved IHS fusion image. Table 1 shows the evaluation result of the fusion image in the city area. It can be seen from the table that improved IHS compared with IHS, NSCT+ improved IHS compared with NSCT + IHS, the average gradient and entropy of the fusion result did not change much, which indicated that the details and edges of the image were well preserved and there is basically no loss of information. Improved IHS compared with HIS, the correlation coefficient and spectral distortion of the fusion result have been improved. The correlation coefficient increased from 0.9 to 0.95 and the spectral distortion decreased from 15.72 to 10.09, which reflects the superiority of this study to improve spectral distortion. NSCT+ improved IHS compared with NSCT + HIS, the correlation coefficient and spectral distortion of the fusion result changed slightly, the correlation coefficient did not change, and the spectral distortion decreased from 7.12 to 6.41, which reflects that the NSCT transform itself can also effectively reduce spectral distortion.

Figure 2 shows the fusion image in the vegetation area, Fig. 2a is panchromatic image, Fig. 2b is multispectral image, Fig. 2c is IHS fusion image, Fig. 2d is improved IHS fusion image, Fig. 2e is NSCT + IHS fusion image, Fig. 2f is NSCT + improved IHS fusion image. Table 2 shows the evaluation results of the fusion image in the vegetation area. It can be seen from the table that the average gradient and entropy of the fusion results of the four algorithms are basically consistent with the urban area, and the change is small and slightly reduced.

This reduction is not a systematic detail and loss of information caused by the improved algorithms. Because the improved algorithm matches the simulated panchromatic image with the I component histogram, the detail retention and the amount of information do not change theoretically, and the specific value of the change is related to the adoption data and the selection area, and the randomness is strong. In contrast, the correlation coefficient and spectral distortion of the fusion result are more effective in the vegetation area, the improved IHS fusion results increased the correlation coefficient from 0.57 to 0.92 for IHS fusion results, and

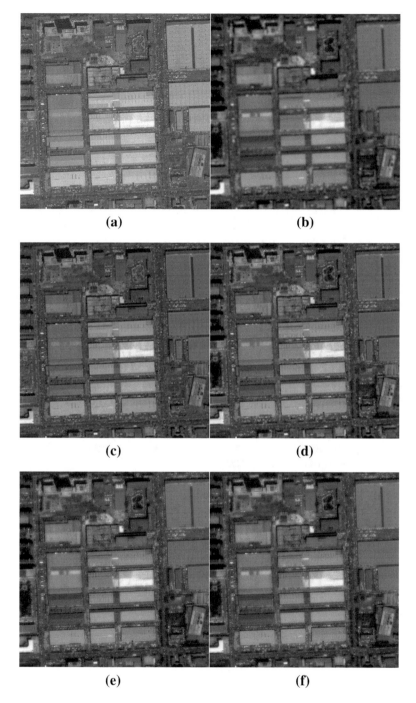

Fig. 1 Fusion image in a city area. (**a**) Panchromatic image. (**b**) Multispectral image. (**c**) IHS fusion image. (**d**) Improved IHS fusion image. (**e**) NSCT + IHS fusion image. (**f**) NSCT + improved IHS fusion image

Table 1 Evaluation of image fusion in A city area

Index		Fusion method			
		IHS	Improved IHS	NSCT + IHS	NSCT + improved IHS
A_G	R	6.75	6.31	6.80	6.34
	G	6.70	6.20	6.61	6.13
	B	7.12	6.55	6.83	6.37
	Ave	6.86	6.35	6.74	6.28
E	R	7.00	7.19	7.30	7.28
	G	6.91	7.00	7.08	7.06
	B	7.12	7.38	7.52	7.50
	Ave	7.01	7.19	7.30	7.28
C_C	R	0.93	0.97	0.98	0.98
	G	0.82	0.91	0.94	0.95
	B	0.95	0.98	0.98	0.98
	Ave	0.90	0.95	0.97	0.97
S_D	R	15.72	10.09	7.12	6.41
	G	15.72	10.09	7.12	6.41
	B	15.72	10.09	7.12	6.41
	Ave	15.72	10.09	7.12	6.41
S_I		2.75	4.30	6.68	6.93

the spectral distortion decreased from 18.68 to 7.08 and the correlation coefficient of NSCT+ improved IHS fusion results also increased from 0.93 to 0.96, and the spectral distortion decreased from 5.68 to 4.10. It can also be seen from Fig. 2 that the improved algorithm results are superior to the IHS in color gradation. This indicates that for the vegetation area, the improved algorithm proposed in this research can further preserve the spectral information based on the effective reduction of spectral distortion by NSCT.

Comparing the fusion results of the two groups of images, the improved fusion method can improve the image with vegetation information more obviously. The comprehensive index value of improved IHS fusion results in urban areas increased from 2.75 to 4.30, and NSCT + improved IHS fusion results increased from 6.68 to 6.93, while improved IHS fusion results in vegetation areas increased from 1.10 to 3.80, and NSCT + improved IHS fusion results increased from 6.00 to 7.46. The reason for the obvious improvement of vegetation area is that the spectral range of panchromatic image has near-infrared spectral bands than three spectral bands involved in fusion. If the brightness of surface objects in near-infrared spectral bands is similar to the other three spectral bands, the improvement of spectral distortion by the improved algorithm is limited. However, the brightness of vegetation in near-infrared spectral bands is obviously higher than that in other three spectral bands, when using IHS fusion without separating near-infrared information in panchromatic images, spectral distortion will occur; this is also the theoretical basis of the improved algorithm proposed in this study.

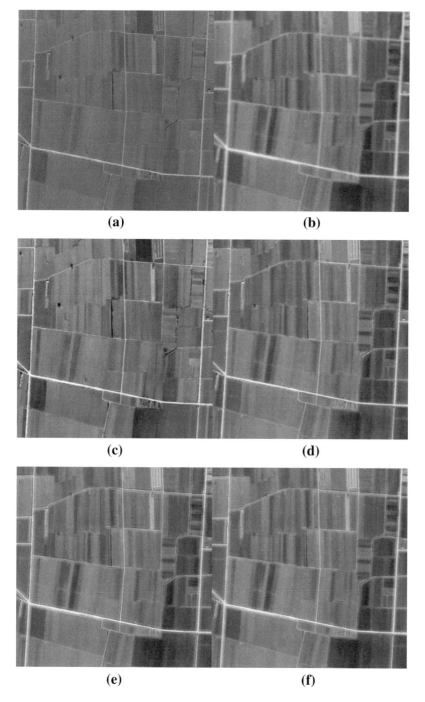

(a) **(b)**

(c) **(d)**

(e) **(f)**

Fig. 2 Fusion image in vegetation area. (**a**) Panchromatic image. (**b**) Multispectral image. (**c**) IHS fusion image. (**d**) Improved IHS fusion image. (**e**) NSCT + IHS fusion image. (**f**) NSCT + improved IHS fusion image

Table 2 Evaluation of image fusion in vegetation area

Index		Fusion method			
		IHS	Improved IHS	NSCT + IHS	NSCT + improved IHS
A_G	R	5.51	4.51	5.53	4.88
	G	5.49	4.44	5.46	4.79
	B	5.37	4.31	5.37	4.67
	Ave	5.46	4.42	5.46	4.78
E	R	6.40	6.55	6.64	6.56
	G	6.48	6.56	6.67	6.62
	B	6.49	6.81	6.89	6.86
	Ave	6.46	6.64	6.73	6.68
C_C	R	0.56	0.91	0.92	0.95
	G	0.56	0.91	0.92	0.95
	B	0.63	0.93	0.94	0.96
	Ave	0.58	0.92	0.93	0.96
S_D	R	18.68	7.08	5.68	4.10
	G	18.68	7.08	5.68	4.10
	B	18.68	7.08	5.68	4.10
	Ave	18.68	7.08	5.68	4.10
S_I		1.10	3.80	6.00	7.46

4 Conclusion

High resolution with rich spectral remote sensing images can be used in low-altitude security monitoring and spatial visualization. This study takes the GF-2 satellite remote sensing image as an example; aiming to solve the problem of spectral distortion caused by the traditional IHS transform fusion algorithm, an improved IHS algorithm is proposed. The high-resolution panchromatic image is reconstructed based on the spectral features of remote sensing image, and to improve the fidelity of fusion image by combining with NSCT transform. Four fusion methods were used to verify the results. The improved IHS raised the comprehensive index value from 2.75 to 4.30, and the vegetation area from 1.10 to 3.80; NSCT + improved IHS raised the comprehensive index value from 6.68 to 6.93 in urban areas, and in the vegetation area from 6.00 to 7.46. Thus, it draws the conclusion that: (1) Compared with the traditional IHS method, the improved IHS method has almost no difference in image details and information quantity, but has a great improvement in spectral information preservation; (2) NSCT transform can effectively reduce spectral distortion, and combining with improved IHS algorithm, the fusion effect can be further improved on the basis of NSCT; (3) The improved algorithm significantly improves the effect of vegetation area than non-vegetation area.

References

1. J. Liu, M. Yong, Y. Wu, F. Chen, Review of methods and applications of high spatiotemporal fusion of remote sensing data. J. Remote Sens. **20**, 1038–1049 (2016)
2. B. Huang, Y. Zhao, Research status and prospect of spatiotemporal fusion of multi-source satellite remote sensing imagery. Acta Geod. Cartogr. Sin. **46**(10), 1492–1499 (2017)
3. X. Yuan, X.J. Yuan, Fusion of multi-planar images for improved three-dimensional object reconstruction. Comput. Med. Imaging Graph. **35**(5), 373–382 (2011)
4. Y. Yaermaimaiti, L.-r. Xie, J. Kong, Remote sensing image fusion based on PCA transform and wavelet transform. Infrared Laser Eng. **43**(7), 2335–2340 (2014)
5. C. Pohl, G.J.L. Van, Multisensor image fusion in remote sensing: concepts, methods and applications. Int. J. Remote Sens. **19**(5), 823–854 (1998)
6. T.M. Tu, S.C. Su, H.C. Shyu, et al., A new look at IHS-like image fusion methods. Inf. Fusion **2**(3), 177–186 (2001)
7. T.M. Tu, P.S. Huang, C.L. Hung, et al., A fast intensity-hue-saturation fusion technique with spectral adjustment for IKONOS imagery. IEEE Geosci. Remote Sens. Lett. **1**(4), 309–312 (2004)
8. M. Choi, A new intensity-hue-saturation fusion approach to image fusion with a tradeoff parameter. IEEE Trans. Geosci. Remote Sens. **44**(6), 1672–1682 (2006)
9. A.-C. Wang, S.-C. Chen, X.-L. Wang, Adaptive remote sensing image fusion based on IHS and wavelet transform. Opto-Electron. Eng. **43**(8), 76–83 (2016)
10. H.-C. Xiao, Q. Zhou, X.-S. Zheng, A fusion method of satellite remote sensing image based on IHS transform and curvelet transform. J. South Chin. Univ. Technol. **44**(1), 58–64 (2016)
11. H.-Y. Cai, L.-R. Zhuo, P. Zhu, et al., Fusion of infrared and visible images based on non-subsampled contourlet transform and intuitionistic fuzzy set. Acta Photon. Sin. **47**(6), 0610002 (2018)
12. P.-D. Yun, Research on pixel level remote sensing image fusion method. Central South University, 2007
13. L. Zhang, S.-L. Han, et al., Fusion of infrared and visual images based on non-sampled contourlet transform and region classification. Opt. Precis. Eng. **23**(3), 810–818 (2015)
14. W. Yi, Y. Zeng, Z. Yuan, Fusion of GF-3 SAR and optical images based on nonsubsampled contourlet transform. Acta Opt. Sin. **38**(11), 1110002 (2018)
15. M.-H. Zhang, H. Chen, W.-Q. Zang, et al., Research of improved IHS image fusion method based on spectral range. J. Henan Univ. **47**(3), 317–322 (2017)
16. C.A. Da, J. Zhou, M.N. Do, The nonsubsampled contourlet transform: theory, design, and applications. IEEE Trans. Image Process. **15**(10), 3089–3109 (2006)
17. K. Shankar, M. Elhoseny, R. Satheesh Kumar, S.K. Lakshmanaprabu, X. Yuan, Secret image sharing scheme with encrypted shadow images using optimal homomorphic encryption technique. J. Ambient Intell. Humaniz. Comput. 2018. https://doi.org/10.1007/s12652-018-1161-0
18. X. Yuan, D. Li, D. Mohapatra, M. Elhoseny, Automatic removal of complex shadows from indoor videos using transfer learning and dynamic thresholding. Comput. Electr. Eng. 2017. https://doi.org/10.1016/j.compeleceng.2017.12.026
19. N. Krishnaraj, M. Elhoseny, M. Thenmozhi, M.M. Selim, K. Shankar, Deep learning model for real-time image compression in Internet of Underwater Things (IoUT). J. Real-Time Image Process. 2019. https://doi.org/10.1007/s11554-019-00879-6
20. M. Elhoseny, K. Shankar, S.K. Lakshmanaprabu, A. Maseleno, N. Arunkumar, Hybrid optimization with cryptography encryption for medical image security in Internet of Things. Neural Comput. Appl. 2018. https://doi.org/10.1007/s00521-018-3801-x
21. K. Shankar, M. Elhoseny, S.K. Lakshmanaprabu, M. Ilayaraja, R.M. Vidhyavathi, M. Alkhambashi, Optimal feature level fusion based ANFIS classifier for brain MRI image classification. Concurrency Comput. Pract. Exp. 2018. https://doi.org/10.1002/cpe.4887

A Unified Coherent-Incoherent Target Decomposition Method for Polarimetric SAR

Shuai Yang, Xiuguo Liu, Xiaohui Yuan, Qihao Chen, and Shengwu Tong

Abstract Polarimetric SAR decomposition is an effective tool to extract ground scattering components and it was dedicated to coherent or incoherent situations. However, coherent targets always mix with incoherent targets in a real situation, which makes decomposition a challenging task. To address this issue, a unified coherent-incoherent decomposition method based on target coherence parameter is proposed. The scattering matrices are transformed into coherency matrices to consolidate decomposition targets. Then a coherency parameter was raised to clarify the coherency of a given target, and decomposition results were generated by combining Krogager and Yamaguchi components via the discrimination of target coherency. Experimental results confirmed that the proposed target coherence parameter is able to differentiate coherent situations from incoherent situations. The unified decomposition can recognize ground scattering and achieve a classification accuracy of 84.96%. Our experimental results reveal that there is still space to improve the performance of unified decomposition.

Keywords Polarimetric synthetic aperture radar · Target decomposition · Coherence

S. Yang
Department of Computer Science and Engineering, University of North Texas, Denton, TX, USA

Faculty of Information Engineering, China University of Geosciences, Wuhan, Hubei, China

X. Liu · Q. Chen · S. Tong
Faculty of Information Engineering, China University of Geosciences, Wuhan, Hubei, China

X. Yuan (✉)
Department of Computer Science and Engineering, University of North Texas, Denton, TX, USA
e-mail: xiaohui.yuan@unt.edu

© Springer Nature Switzerland AG 2020
X. Yuan, M. Elhoseny (eds.), *Urban Intelligence and Applications*, Studies in
Distributed Intelligence, https://doi.org/10.1007/978-3-030-45099-1_6

1   Introduction

Polarimetric Synthetic Aperture Radar (PolSAR) has drawn great attention from researchers in the past decades because the data can be acquired under all-day/all-weather conditions [1, 2] for a variety of applications such as vegetation monitoring [3] and land cover delineation [4]. Target decomposition is one of the most effective ways to analyze scattering properties[5, 6], which can be categorized into two groups. The first is coherent decomposition for the scattering matrix, and the other category is incoherent decomposition for coherency/covariance matrix. Coherent decomposition requires determinate ground objects with coherent back-scattering, and on average is needed before decomposition. Several classical coherent decompositions were proposed: including Pauli decomposition [7], Krogager decomposition [8], Cameron decomposition [9], and Touzi decomposition [10]. As for incoherent decomposition, the ground objects can be indeterminate, i.e., changing with time or space, and the back-scattering can be incoherent or partially coherent. Some typical incoherent decomposition methods include Cloude decomposition [11], Freeman decomposition [12], and Yamaguchi decomposition [13].

Coherent decomposition is widely employed for processing high resolution and low entropy scattering issues, and it ignores the severe impact of speckle noise in the single-look PolSAR images. On the other hand, incoherent decomposition is applicable to a large number of non-deterministic targets in nature. With the increased resolution and range of PolSAR, the acquired data are more likely to have both coherent and incoherent objects in the real world. The existing decomposition methods deal with either coherent or incoherent situations, which is ineffective for processing many PolSAR data and a unified decomposition is hence needed.

To handle both coherent and incoherent decomposition seamlessly, we propose a unified coherent-incoherent decomposition based on the coherence degree of the targets. In our method, the coherence degree is computed for each pixel, which is decided as a coherent or an incoherent pixel based on a threshold. Krogager decomposition and Yamaguchi decomposition are performed to the input data. The final results of the unified decomposition are obtained by fusing the decomposition from Krogager and Yamaguchi methods based on the category of each pixel.

The rest of this paper is organized as follows: Sect. 2 gives details of the proposed target coherence parameter and the unified coherent-incoherent decomposition. Section 3 discusses our experimental results using ESAR and RADARSAT-2. This paper is concluded in Sect. 4 with a summary and future work.

2   Unified Coherent-Incoherent Decomposition

2.1   PolSAR Data and Decomposition

The PolSAR data is represented by the polarimetric scattering matrix, i.e.,

$$S = \begin{bmatrix} S_{HH} & S_{HV} \\ S_{HV} & S_{VV} \end{bmatrix}, \tag{1}$$

where subscripts H and V represent horizontal and vertical direction, respectively. When reciprocity holds, we have $S_{HV} = S_{VH}$. The scattering matrix only adapts coherent targets. To characterize indeterminate objects, Pauli scattering vector k and coherency matrix T are introduced:

$$k = \frac{1}{\sqrt{2}}[S_{HH} + S_{VV} \quad S_{HH} - S_{VV} \quad 2S_{HV}]', \tag{2}$$

$$T = \frac{1}{L} \sum_{i=1}^{L} k_i k_i^H, \tag{3}$$

where k^H is the conjugate transpose of k, the superscript $[\cdot]'$ means transpose.

Among coherent decompositions, Krogager is the most relative to incoherent decompositions as it decomposes scattering matrix into three components with physical meanings, includes sphere, double bounce, and helix scattering. The Krogager decomposition is expressed as:

$$S = e^{j\phi} \left(e^{j\phi_s} k_s \begin{bmatrix} 1 & 0 \\ 0 & 1 \end{bmatrix} + k_d \begin{bmatrix} \cos 2\theta & \sin 2\theta \\ \sin 2\theta & -\cos 2\theta \end{bmatrix} \right.$$
$$\left. + k_h e^{\mp j2\theta} \begin{bmatrix} 1 & \pm j \\ \pm j & -1 \end{bmatrix} \right), \tag{4}$$

where ϕ denotes the absolute phase of scattering matrix, k_s, k_d, and k_h represent sphere, double bounce, and helix components, respectively. The latter two components contain a spin angle θ.

As for incoherent decomposition, the most accepted method is Yamaguchi decomposition. The coherency matrix is divided into four components f_s, f_d, f_v, and f_h, which denotes surface, double bounce, volume, and helix scattering, respectively:

$$T_3 = \frac{f_s}{1 + |\beta|^2} \begin{bmatrix} 1 & \beta^H & 0 \\ \beta & |\beta|^2 & 0 \\ 0 & 0 & 0 \end{bmatrix} + \frac{f_d}{1 + |\alpha|^2} \begin{bmatrix} |\alpha|^2 & \alpha & 0 \\ \alpha^H & 1 & 0 \\ 0 & 0 & 0 \end{bmatrix}$$
$$+ \frac{f_v}{4} \begin{bmatrix} 2 & 0 & 0 \\ 0 & 1 & 0 \\ 0 & 0 & 1 \end{bmatrix} + \frac{f_h}{2} \begin{bmatrix} 0 & 0 & 0 \\ 0 & 1 & \pm j \\ 0 & \mp j & 1 \end{bmatrix}, \tag{5}$$

where α and β are the complex coefficient of double bounce and volume scattering, respectively. Yamaguchi modified this decomposition by adding de-orientation process and revising the volume model, which reduces the over-estimated volume component and eliminates the negative power.

It is obvious that the results of Krogager decomposition and Yamaguchi decomposition have a good correspondence in terms of the scattering type. The sphere, double bounce, and helix components of Krogager decomposition match surface, double bounce, and helix scattering, respectively.

2.2 Coherence Parameter

The key point of unified coherent-incoherent decomposition is to clarify the coherence of pixel since coherent decomposition suits for coherent targets while incoherent decomposition fits incoherent decomposition better.

Touzi and Charbonnequ [10] introduced the signal-to-clutter ratio to clarify the coherence. The scattering matrix is decomposed with a Huynen vector to obtain eigenvalues:

$$
\begin{cases}
\lambda_1 = m e^{j2v} \\
\lambda_2 = m \tan^2(\gamma)\, e^{-j2v}
\end{cases}
\tag{6}
$$

where λ_1 and λ_2 are the largest and smallest eigenvalues of the scattering matrix. m^2 is the largest polarimetric return. Target with signal-to-clutter higher than 15 dB can be considered as coherent. However since the calculation of signal-to-clutter requires scattering matrix, which does not match the coherency matrix in incoherent decomposition.

It is known that coherent pixel contains only one scattering component theoretically. So the dominant scattering can be used to clarify the coherence of pixel. Divide the coherency matrix into three parts, i.e., eigenvalue decomposition.

$$
[T] = \lambda_1 e_1 e_1^H + \lambda_2 e_2 e_2^H + \lambda_3 e_3 e_3^H,
\tag{7}
$$

where λ_i is the eigenvalue and e_i denotes the eigenvector. T represents a rank-1 coherency matrix corresponding to one unique scattering. In this vein, λ_1 denotes dominant scattering, while λ_2 and λ_3 denote second scattering and unpolarized components, respectively.

Anisworth [14] pointed out that averaged coherency matrix can be divided into polarized and unpolarized terms, and the polarization fraction (PF) is further defined as

$$
PF = 1 - \frac{3\lambda_3}{\lambda_1 + \lambda_2 + \lambda_3}.
\tag{8}
$$

We propose two novel metrics to clarify the coherence of the target. The first is Maximum Eigenvalue Fraction (MEF), i.e., the ratio of the largest eigenvalue and span power:

$$MEF = \frac{\lambda_1}{\lambda_1 + \lambda_2 + \lambda_3}, \tag{9}$$

another metric is defined as target coherence parameter (TC):

$$TC = 1 - \sqrt{\frac{3}{2}} \sqrt{\frac{\lambda_2^2 + \lambda_3^2}{\lambda_1^2 + \lambda_2^2 + \lambda_3^2}}. \tag{10}$$

2.3 Unified Decomposition

According to the different decomposition objects, the traditional target decomposition can be divided into two types: coherent decomposition and non-coherent decomposition, respectively, for the scattering matrix and the coherent matrix. The scattering matrix can correspond to the basic scattering type of the feature, but it cannot be used to describe instability. In order to solve the coherent non-coherent target mixture in the real world, where the existing decomposition cannot take into account all the problems. The coherent non-coherent unified decomposition is proposed. Based on Eq. (10), the coherence is determined and the appropriate decomposition method is selected for decomposition. The algorithm is demonstrated as follows:

Algorithm 1 Unified decomposition algorithm

1: Coherency calculation and division: Transform the scattering matrix to the coherency matrix, calculate the Target Coherence parameter, then divide the range into coherent and incoherent based on a given threshold.
2: Coherent decomposition: For coherent pixels, use Krogager decomposition and compute the percentage of each component, keep the dominant scattering and let it equals the total power of the current pixel.
3: Incoherent decomposition: For incoherent pixels, use Yamaguchi decomposition.
4: Final results generation: Combine two results into one image, i.e., coherent pixels with Krogager decomposition and incoherent pixels with Yamaguchi decomposition. Generalize the results to obtain the final results.

---

3 Experiments

A dataset used in our experiments is an L band ESAR data of Oberpfaffenhofen, Germany. As shown in Fig. 1, the size is 2816×1540 with a resolution of $1.5\,m \times 0.89\,m$. Another dataset is a C band RADARSAT-2 of Flevoland, Holland.

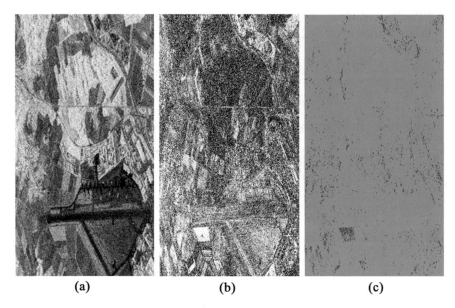

(a)   (b)   (c)

Fig. 1 PauliRGB of ESAR in Oberpfaffenhofen and its target coherence parameter and its division. (**a**) PauliRGB. (**b**) Target coherence. (**c**) Division

As shown in Fig. 2, the size is 800×1540 with a resolution of approximately $8\,\text{m} \times 8\,\text{m}$.

The key point of unified decomposition is the accuracy of the target coherence metric. PF, MEF, and TC are all calculated from eigenvalues. Hence it is straightforward to compare these metrics with different eigenvalues. Hence we select several groups and generalize them to meet requirement of $\lambda_1 + \lambda_2 + \lambda_3 = 1$.

As shown in Table 1, when in the complete coherent situation. all three methods obtained the highest value, i.e., 1. As for the complete incoherent situation, all three methods contain the smallest value. It is obvious that they are able to characterize complete coherent and incoherent targets. For samples 2 and 3, the targets are no longer completely coherent, but PF still got the highest value which shows not a valid evaluation of target coherence. MEF can characterize most situations, but it fails to distinguish 3 from 5. TC is the only metric that can characterize all six situations with success.

The results of unified decomposition are further carried out with ESAR data and RADATSAT-2 data, as demonstrated in Figs. 3 and 4, respectively. It is clear that unified decomposition is able to obtain reasonable results. For traditional incoherent decompositions, volume scattering is always over-estimated. We compared the averaged volume component percentage as shown in Table 2. It is clear that volume scattering is reduced compared to Yamaguchi decomposition. This result is rational since none volume scattering exists in real-world coherent targets.

We use decomposition components as the input features to perform classification. The results are demonstrated in Fig. 5, where higher accuracy of classification means

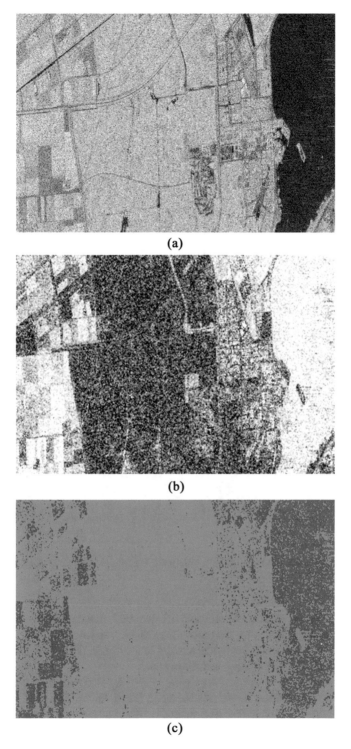

Fig. 2 PauliRGB of RADARSAT-2 in Flevoland and corresponding figures of target coherence parameter and its division. (**a**) PauliRGB. (**b**) Target coherence. (**c**) Division

Table 1 Comparison of coherence parameters with different eigenvalues

Sample	λ_1	λ_2	λ_3	PF	MEF	TC
1	1	0	0	1	1	1
2	0.75	0.25	0	1	0.75	0.61
3	0.5	0.5	0	1	0.5	0.13
4	0.5	0.4	0.1	0.5	0.7	0.22
5	0.5	0.25	0.25	0.25	0.5	0.29
6	0.33	0.33	0.33	0	0.33	0

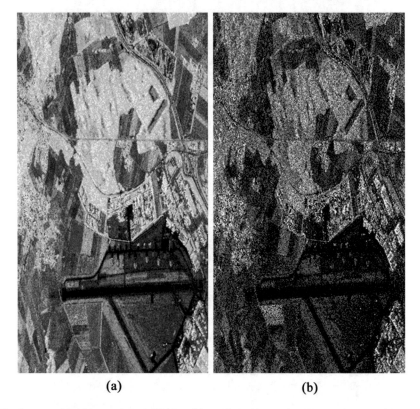

(a)   (b)

Fig. 3 Decomposition results of ESAR with unified coherent-incoherent decomposition. (**a**) Combined RGB. (**b**) Helix component

better characterization of Polarimetric SAR data with decomposition components. As a comparison, coherent decomposition-based classification and incoherent decomposition-based classification is marked in the figure. It is clear that incoherent decomposition obtained better accuracy overall.

The ideal classification accuracy is calculated by the fusion of coherent and incoherent decomposition results. As shown in Fig. 5, the upper bound accuracy of the combination is much higher than that of coherent or incoherent decomposition. It is clear that our method still has room to improve.

Fig. 4 Decomposition results of RADARSAT-2 with unified coherent-incoherent decomposition. (**a**) Combined RGB. (**b**) Helix component

Table 2 Comparisons of volume component percentage with different decompositions

	Yamaguchi decomposition	Unified decomposition
ESAR	40.2%	39.8%
RADARSAT-2	40.9%	38.4%

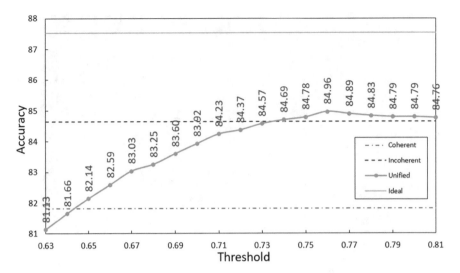

Fig. 5 Classification accuracy of RADARSAT-2 with different decomposition

With the threshold increases in unified coherent-incoherent decomposition, the accuracy increases as well. When threshold exceeds 0.73, the accuracy of unified decomposition surpasses that of incoherent decomposition, which confirms that unified coherent-incoherent decomposition characterizes ground objects better compared to coherent decomposition and incoherent decomposition. These results reveal that unified decomposition with a proper threshold can better describe ground types.

4 Conclusion

This paper presents a unified coherent-incoherent decomposition method based on a coherence parameter. Target coherence is calculated to clarify the coherency of each pixel. Then the pixels are categorized into coherent and incoherent ones. Krogager decomposition is employed for processing the coherent parts, and Yamaguchi decomposition is employed for processing the incoherent parts. The results are obtained by combining coherent and incoherent components into unified decomposition components.

Our experimental results demonstrate that the proposed decomposition method is a favorable and promising way to extract ground object information from PolSAR data. Classification using our unified coherent-incoherent decomposition obtained better accuracy with a proper threshold compared to the state-of-the-art coherent and incoherent decomposition methods. On the other hand, there is still much room for improvement.

Acknowledgments This work was supported by the National Natural Science Foundation of China under Grant No. 41471355 and 41771467 and funds by the China Scholarship Council (CSC) during a visit of Shuai Yang at the University of North Texas.

References

1. H. Aghababaee, M.R. Sahebi, Incoherent target scattering decomposition of polarimetric SAR data based on vector model roll-invariant parameters. IEEE Trans. Geosci. Remote Sens. **54**(8), 4392–4401 (2016)
2. X. Yuan, I. Ternovskiy, Image reconstruction from sub-apertures of circular spotlight SAR, in *SPIE Defense + Security*, Baltimore, MD, vol. 9458 (2015)
3. D. Haldar, R. Dave, V.A. Dave, Evaluation of full-polarimetric parameters for vegetation monitoring in rabi (winter) season. Egypt. J. Remote Sens. Space Sci. **21**, S67–S73 (2018). EJRS Special Issue: Microwave Remote Sensing in honor of Professor Adel Yehia
4. X. Yuan, V. Sarma, Automatic urban water-body detection and segmentation from sparse ALSM data via spatially constrained model-driven clustering. IEEE Geosci. Remote Sens. Lett. **8**(1), 73–77 (2010)
5. X. Wang, L. Zhang, S. Zhu, A four-component decomposition model for polarimetric SAR images based on adaptive volume scattering model, in *IGARSS 2018 – 2018 IEEE International Geoscience and Remote Sensing Symposium*, July 2018, pp. 4563–4566
6. S. Yang, Q. Chen, X. Yuan, X. Liu, Adaptive coherency matrix estimation for polarimetric SAR imagery based on local heterogeneity coefficients. IEEE Trans. Geosci. Remote Sens. **54**(11), 6732–6745 (2016)
7. J.-S. Lee, E. Pottier, *Polarimetric Radar Imaging: From Basics to Applications* (CRC Press, Boca Raton, 2009)
8. E. Krogager, New decomposition of the radar target scattering matrix. Electron. Lett. **26**(18), 1525–1527 (1990)
9. W.L. Cameron, L.K. Leung, Feature motivated polarization scattering matrix decomposition, in *Record of the IEEE 1990 International Radar Conference, 1990* (IEEE, New York, 1990), pp. 549–557
10. R. Touzi, F. Charbonneau, Characterization of target symmetric scattering using polarimetric sars. IEEE Trans. Geosci. Remote Sens. **40**(11), 2507–2516 (2002)
11. S.R. Cloude, E. Pottier, An entropy based classification scheme for land applications of polarimetric SAR. IEEE Trans. Geosci. Remote Sens. **35**(1), 68–78 (1997)
12. A. Freeman, S.L. Durden, A three-component scattering model for polarimetric SAR data. IEEE Trans. Geosci. Remote Sens. **36**(3), 963–973 (1998)
13. Y. Yamaguchi, A. Sato, W.-M. Boerner, R. Sato, H. Yamada, Four-component scattering power decomposition with rotation of coherency matrix. IEEE Trans. Geosci. Remote Sens. **49**(6), 2251–2258 (2011)
14. T.L. Ainsworth, J.S. Lee, D.L. Schuler, Multi-frequency polarimetric SAR data analysis of ocean surface features, in *Geoscience and Remote Sensing Symposium, 2000. Proceedings. IGARSS 2000. IEEE 2000 International*, vol. 3 (IEEE, New York, 2000), pp. 1113–1115

Part II
Community and Wellbeing of Smart Cities

Urban Land Use Classification Using Street View Images Based on Deep Transfer Network

Yafang Yu, Fang Fang, Yuanyuan Liu, Shengwen Li, and Zhongwen Luo

Abstract Urban land use maps are significant references for urban planning and environmental research. Contrary to land cover mapping, it is generally impossible using overhead imagery since it requires close-up views. Street view images capture the surrounding scenes along streets and represent urban land information objectively. In this work, we proposed an automatic classification model using street view images for urban land use classification. We utilize high-level semantic image features extracted adopting a deep transfer network, which includes three fully-connected layers. Geographic information was applied to mask out land use parcel and to associate the corresponding street view images. The approach allows us to achieve 61.8% accuracy on a challenging six class land use mapping problem. The assessment results show that the proposed approach has potential on land use classification. Since the street view images are publicly accessible and supply a variety of APIs for downloading, the presented approach would provide an effective way for urban-related research in future.

Keywords Land use · Transfer learning · Street view images · Classification

1 Introduction

A detailed and accurate urban land use map contains a wealth of information, which can provide a reference for urban planning and environmental monitoring [1, 2]. However, land use maps are usually created by field surveys, which consume enormous human efforts and financial resources. Hence, it is of great significance to establish a model that automatically generates accurate and up-to-date land use maps.

Y. Yu · F. Fang (✉) · Y. Liu · S. Li · Z. Luo
College of Information Engineering, China University of Geosciences, Wuhan, China
e-mail: fangfang@cug.edu.cn

© Springer Nature Switzerland AG 2020
X. Yuan, M. Elhoseny (eds.), *Urban Intelligence and Applications*, Studies in Distributed Intelligence, https://doi.org/10.1007/978-3-030-45099-1_7

Many existing methods are mainly based on high-resolution remote sensing images [3–8]. Although it might be easy to distinguish airports and residential areas using overhead imagery [9], it is much more difficult to determine land use in complicate urban areas. Images taken at ground-level are potentially more indicative. The ground-level image acquisition is easy and the data are made available in many online services [10–12]. These data sources derived from geo-crowdsourcing and applications of volunteered geographic information (VGI), such as photos from Instagram, Panoramio, and Flickr, which are large and informative. However, such images are mostly distributed unevenly in space and contain a great amount of noise.

Street view images are informative and provide us a significant data source for urban studies from a different perspective. It has shown great potential for deriving urban environment information [13–15], for instance, a deep learning approach to assess the quality of urban visual environment [16], image feature descriptors to extract building-based land use information [17], aerial and street view images to a fine-grained catalog of street trees [18]. While street view images not only include basic information such as the geographic location of the image but also contain rich scene semantic of urban land use. Therefore, based on scene semantic of the street view images and combined with the related geographic vector data, we propose a method to automatically obtain the urban land use map. A deep transfer learning method is adopted to extract image features. In order to reduce street view image's geographic location positioning errors, we used shape files to filter photos.

The rest of the paper is organized as follows. We present the details of our land use classification framework and methods in Sect. 2. In Sect. 3, we describe our study area and dataset, experimental setup, results, and geo-visualizations. Section 4 concludes this paper and presents the future work of our research.

2   Method

The framework of the proposed land use classification method is shown in Fig. 1. Tencent Street View (TSV) was utilized as the image dataset. The method is split into three processes. Firstly, TSV images of land use along streets were collected and labeled based on the existing land use maps, then geo-filtering with shape files. Our method does not require individual TSV images to be manually labeled. Secondly, a classification model [25, 26] was built, which uses deep transfer network to automatically classify the land use. The probabilities of the land use classes are calculated in the model and are considered as the weights of classification in each photo. Lastly, the mapping set using a trained three fully-connected layers was applied to test TSV images classification result in the study area. And a strategy of determining land use types was used to assess the accuracy of land use maps.

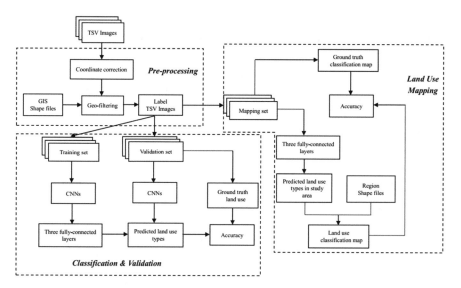

Fig. 1 The framework of the proposed method using TSV images. The workflow includes a diagram of image processing, major classification procedures, and land use mapping

2.1 Image Harvesting and Pre-processing

The TSV images are taken along the street to capture land use parcel information which is not directly related to it. Therefore, it is necessary to associate TSV images with the land use parcel in order to obtain labels. The details are as follows: we downloaded TSV images for different heading angles (the heading angles range from 0° to 360°) from the TSV Static Image API [1]. The camera used to take TSV images is 0° in the vertical direction, so as to ensure street view information is captured. The horizontal direction is set to 0°, 90°, 180°, and 270°, as shown in Fig. 2. According to the heading angles of the obtained TSV images, we set a threshold to control the distance from TSV images to land use. In this way, the adjusted geographical coordinates, (Gxi, Gyi), fall within or above the land use parcel, as defined follows:

$$(Gxi, Gyi) = \begin{cases} (Gx, Gy + m), \text{angle} = 0° \\ (Gx + m, Gy), \text{angle} = 90° \\ (Gx, Gy - m), \text{angle} = 180° \\ (Gx - m, Gy), \text{angle} = 270° \end{cases}, \quad i = 1, \ldots, 4 \tag{1}$$

where (Gx, Gy) represents the geographic coordinates of the downloaded TSV images, m is a variable threshold indicating the width of the street where the TSV imagery is located, which sets to 80 m in our study. The angle from 0° to 270° corresponds to the north, east, south, and west in the geographic coordinate

Fig. 2 The left side is a picture taken from four angles of a TSV image (a yellow point in the picture), and the right side is the geographic coordinate system orientation corresponding to each angle

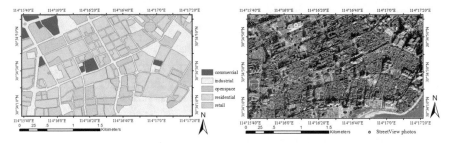

Fig. 3 Region shape files from OpenStreetMap (left) and distribution of TSV images in the study area (right)

system. In this way, we associated the adjusted TSV images with land use parcel and get the final label for all TSV images.

We use the polygonal outlines of the land use areas to filter the noise imagery and to produce a more precise land use map. These irregularly shaped polygons are known as shape files from OpenStreetMap [2] and are widely available. There are two advantages to use the shape files shown in Fig. 3 (Left): (1) Filtering: We ignore photos that are not within the area we want to classify. This eliminates many noisy (unrelated) photos and reduces our dataset from 23,690 photos to 8059 photos; (2) Accuracy: The land use map was generated using the shape files, just very few modifications can be used. We use the corresponding shape files to compare the experimental results.

2.2 A Deep Transfer Network

A deep transfer network [19] is used to fine-tune a pre-trained model so that the complexity of the model in the experiment is reduced. A pre-trained deep network includes model structure and network weights for large datasets training. In this

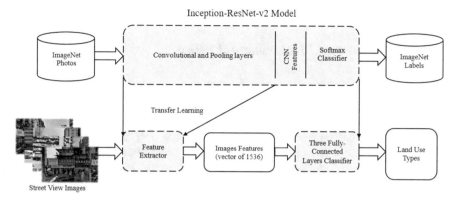

Fig. 4 A deep transfer network of this study

study, the Inception-ResNet-v2 model [20] from Google was pre-trained by the ImageNet dataset [21] is adopted. A deep transfer network [22–24] of our study is shown in Fig. 4. By removing the last output layer of the pre-trained Inception-ResNet-v2 model, all the convolutional and pooling layers are taken out as the feature extractor with the output of 1536 CNN codes (image features). The CNN codes are classified by a model with three fully-connected layers for land use type recognition. The three fully-connected layers model includes 1024, 256, and 256 CNN nodes, respectively. And the last layer of the fully-connected layers is reduced from the pre-trained model 1792 nodes [20] to our model 256 nodes.

2.3   A Strategy of Determining Land Use Types

After the network training and verification are completed, we have an optimized classifier for image classification. To improve the robustness, the mapping set using the classifier, a trained three fully-connected layers, is applied to test TSV images classification result in study area, and the land use classes are obtained at a decision level. We apply a rule to assess the accuracy of land use maps. The rule used to determine the building instances used in [14] is referenced. Assuming there are N images retrieved of the study area, the final land use class is determined as follows:

$$y_i = \text{argmax} \frac{1}{N} \sum_{k=0}^{N-1} f(i)^k \ (i = 1, 2, \ldots, m) \qquad (2)$$

where $f(i)^k$ is the ith element of three fully-connected layers output f^k in the CNN model and denotes the probability distribution over the whole land use classes, and k is the index of the classified image, i denotes the ith type of image, m represents the total number of land use classes. We obtain the land use class by averaging the probability distribution vectors of all images within a certain parcel.

3   Experiments and Results

3.1   Study Area and Experimental Setup

Figure 5 shows the study area in Wuhan including a variety of land use types. The TSV images were downloaded using the TSV static Image API, and preprocessed using shape files.

Our dataset consists of 8059 TSV images spread over 6 land use categories: commercial, green land, industrial, open space, residential, and retail. Table 1 demonstrates descriptions of the land use classes. The training set consists of 6890 images and we split the dataset with a ratio of 0.8 to 0.2 for training and validation of our model. The image-level classification accuracy of our model is evaluated on the validation set. The mapping set consists of 1169 TSV images as shown in Fig. 3 (right).

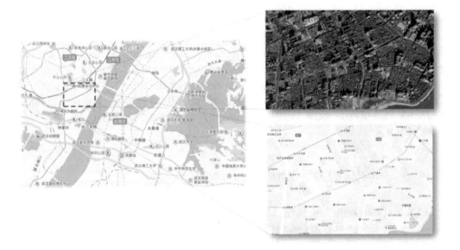

Fig. 5 Location of the study area in Wuhan

Table 1 Land use type descriptions from OpenStreetMap

Land use type	Description
Commercial	Financial services (e.g., banks); service building (e.g., offices building, hotel, market)
Green land	Green land (e.g., forest and grass)
Industrial	Construction areas (e.g., factories, companies, and industrial buildings)
Open space	Open land (e.g., parks and sport fields, square, cemetery)
Residential	Multiple unit or multi-family housing single house for living
Retail	Retail places (e.g., grocery market); mixed function (e.g., shops and apartments)

We use pre-trained model Inception-ResNet-v2. Convolutional layers of the networks are initialized by the pre-trained with ImageNet dataset, and fully-connected layers are randomly initialized. In the fine-tuning, the last layer is a 6-way classifier. The model is trained using stochastic gradient descent algorithm with a learning rate of $\eta = 0.01$ and a momentum value of $p = 0.9$. The batch size is set to 16. To adjust the learning rate, we decay its value by a factor of 0.94 in every 2 epochs. Cross-entropy loss is utilized for training with the weight decay parameter of $w = 4 \times 10^{-5}$. The neurons of fully-connected layers are dropped out by a probability of 5%. All the experiments are implemented with TensorFlow and carry out by one NVIDIA GeForce GTX 1080 Ti 11GBGPU.

3.2 Image Classification

The accuracy of classifying TSV images is evaluated with fivefold cross-validation and the results are shown in Table 2. Among them, producer's accuracy and user's accuracy are related to the confusion matrix. The former refers to the proportion of a certain type of pixel, which is the ratio of the predicted correct category to the total number of actual correct categories of the class. It can be calculated by using the column of the confusion matrix. The latter refers to the proportion of a certain type of pixel, which is predicted as the ratio of the correct category to the total number of predictions of the class. It can be calculated using the rows of the confusion matrix.

In the experiments, the overall accuracy is 73.1% and the kappa value is 0.653. The producer's accuracies of the green land, open space, and residential are 79.3%, 79.0%, and 85.1%, respectively. The user's accuracies corresponding to the three classes are 85.3%, 85.0%, and 67.6%. In our experiments, green land and open space are relatively easier to recognize compared to other classes, while classes, e.g. industrial and retail, are not easily identified. Figure 6 shows the predicted number of TSV images for per-class land use by visualization.

The labels from the OpenStreetMap (OSM) consist of errors among some land use classes, especially for those with similar structures, e.g. some residential and

Table 2 Classification accuracy of TSV images for per-class land use

Land use class	Producer's accuracy (%)	User's accuracy (%)
Commercial	61.4	60.3
Green land	79.3	85.3
Industrial	36.5	69.7
Open space	79.0	85.0
Residential	85.1	67.6
Retail	49.7	74.5
Accuracy	73.1	
Kappa	0.653	

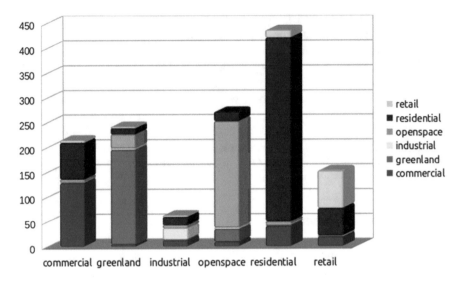

Fig. 6 The predicted number of TSV images for per-class land use. The horizontal axis represents the land use class, and the vertical axis represents the predicted numbers of class

industrial. As shown in Fig. 7 (top left), the land use displayed by the TSV image tends to be a residential, while the land use tag retrieved from OSM is industrial. In addition, the scene semantic of an image is ambiguous. As shown in Fig. 7 (bottom left and top right), in which the above is residential and the below is retail. Lastly, some images include multiple land use classes. As shown in Fig. 7 (bottom right), left is residential and right is commercial. And the land use shown in the figure is commercial, and it is misclassified as a residential.

3.3 Land Use Classification

We label the map regions using the probability distribution vectors of the contained images. We do not attempt to label the regions, which contain no image. In addition, we merge green land and open space into open space class in the final land use map. Since there are few photos of these two types of TSV images in the study area, and there are similar features. Therefore, the final land use class has changed from the original six classes to five classes, including commercial, industrial, open space, residential, and retail.

Our method achieved 59.4% classification accuracy. Table 3 lists the accuracies and the kappa value. Among all land use types, the producer's accuracy for industrial and retail are 39.1% and 54.2%, respectively. This is partially due to a relatively low number of mapping set and a similar structure is with residential.

Industrial (ground truth), residential (predicted)   residential (ground truth), retail (predicted)

retail (ground truth), residential (predicted)   commercial (ground truth), residential (predicted)

Fig. 7 Some results show the reasons that may induce classification errors

Table 3 Classification accuracy of TSV images based on the mapping set

Land use class	Producer's accuracy (%)	User's accuracy (%)
Commercial	55.3	45.5
Green land	100	17.6
Industrial	39.1	52.9
Open space	100	37.5
Residential	63.3	68.4
Retail	54.2	57.9
Accuracy	59.4	
Kappa	0.349	

The region-level predicted result is displayed as a map in Fig. 8 (below). It is reasonable comparing with the classification map (Fig. 8 (above)) from OpenStreetMap, where different colors represent different land use types. We also draw the corresponding confusion matrix of the predicted result in Table 4. The total number of land use in this area is 161. Our result predicts 15 commercial, 2 industrial, 2 openspace, 68 residential, and 49 retails. 25 land use are not classified due to the lack of TSV images. We correctly classify 84 of the remaining 136 for an accuracy of 61.8%. The land use mapping performance is closely related to the TSV image classification accuracy.

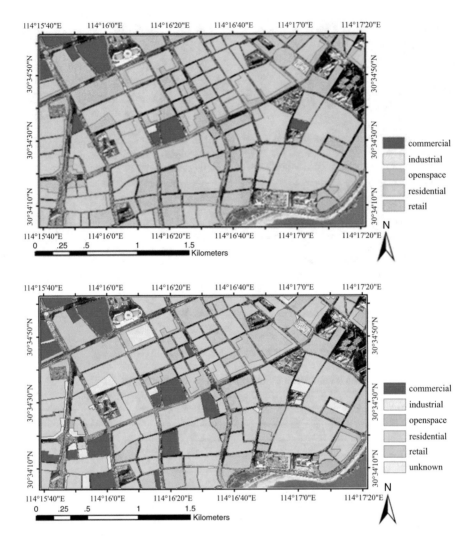

Fig. 8 The predicted land use classification map (below), along with the classification map from OpenStreetMap (above), where different colors represent different land use classes

3.4 Discussion

In this section, we evaluate our proposed method for land use classification. We first compare the results of different pre-trained models. Moreover, we indicate the advantage of the three fully-connected layers classifier in the situation of using TSV images. We also follow the experimental results with a discussion.

The results are shown in Table 5, which are evaluated on two metrics. One is the image classification accuracy, and the other is the land use classification accuracy.

Table 4 The confusion matrix of the classification result of the area in Wuhan

	Commercial	Industrial	Open space	Residential	Retail	Total
Commercial	6	0	0	1	0	7
Industrial	0	0	0	0	0	0
Open space	0	0	2	0	0	2
Residential	6	1	0	49	22	78
Retail	3	1	0	8	27	49
Total	15	2	2	68	49	136
Accuracy		61.8%				

Table 5 The evaluation results of different methods

Method	Image classification		Land use classification	
	Accuracy (%)	Kappa	Test accuracy (%)	Mapping accuracy (%)
AlexNet+three fully-connected layers	77.5	0.713	53.0	53.7
VGG16+three fully-connected layers	73.3	0.656	57.7	58.8
Inception-ResNet-v2+softmax	70.5	0.616	58.7	61.3
Inception-ResNet-v2+three fully-connected layers	73.1	0.653	59.4	61.8

The former is computed based on the keyword labels, while the latter is calculated based on the land use classification map from OpenStreetMap of Wuhan.

Table 5 shows the image classification results for each pre-trained CNN model. Overall accuracies of 77.5%, 73.3%, and 73.1% were calculated based on the AlexNet, VGG16, Inception-ResNet-v2 models, respectively. At the same time, Kappa of them is 0.713, 0.656, and 0.653, respectively. Although the VGG16 and AlexNet models have slightly better performance results based on the image classification accuracy than the model used in this paper, they do not demonstrate a satisfactory accuracy in land use classification. We can see from Table 5 that Inception-ResNet-v2 model shows the best results in land use classification, especially the mapping accuracy is 61.8%. In addition, we display the effectiveness of our proposed three fully-connected layers classifier. It improves nearly 3% over the softmax classifier in image classification. And the mapping accuracy is also improved by 0.5. The above analysis shows that the deep transfer network combined with three fully-connected layers classifiers has a positive impact on the land use classification results.

4 Conclusion

Urban land use mapping is an important basis for a series of applications such as urban planning, environmental research. The rational classification of urban land use helps to grasp the urban spatial structure and formulate scientific development plans for the city. However, traditional methods based on field surveys or remote sensing imagery have some challenges for complex urban land use classification. Therefore, this study proposed a novel approach to apply TSV images for land use classification. We established a deep transfer network with three fully-connected layers model. Our method finally obtained 73.1% accuracy and 0.653 kappa on TSV image-level classification. Based on the 136 land use classifications in the study area, 1169 mapping set achieved 59.4% accuracy. And the urban land use mapping achieved 61.8% accuracy using the strategy of determining land use types. The results above demonstrate that land use classification is possible extracted high-level image features from TSV images using deep transfer network.

Future research will focus on the following aspects. Based on the characteristics of TSV images mainly along the street, the research objects in the images are unclear and unevenly distributed, so we can consider the object-based method for urban land use mapping. Additionally, the feasibility of automatic land use classification with other sources of geo-tagged data will be explored.

References

1. C. Liu, B.H. Henderson, D. Wang, et al., A land use regression application into assessing spatial variation of intra-urban fine particulate matter (PM2.5) and nitrogen dioxide (NO2) concentrations in City of Shanghai, China. Sci. Total Environ. **565**, 607–615 (2016)
2. G.L. Feyisa, H. Meilby, G.D. Jenerette, et al., Locally optimized separability enhancement indices for urban land cover mapping: Exploring thermal environmental consequences of rapid urbanization in Addis Ababa, Ethiopia. Remote Sens. Environ. **175**, 14–31 (2016)
3. D.F. Hong, N. Yokoya, N. Ge, et al., Learnable manifold alignment (LeMA): a semi-supervised cross-modality learning framework for land cover and land use classification. ISPRS J. Photogramm. Remote Sens. **147**, 193–205 (2019)
4. S. Abdikan, F.B. Sanli, M. Ustuner, et al., Land cover mapping using SENTINEL-1 SAR data. ISPRS – Int. Arch. Photogramm. Remote Sens. Spat. Inf. Sci. **7**, 757–761 (2016)
5. G. Suresh, R. Gehrke, T. Wiatr, et al., Synthetic aperture radar (SAR) based classifiers for land applications in Germany. ISPRS – Int. Arch. Photogramm. Remote Sens. Spat. Inf. Sci. **1**, 1187–1193 (2016)
6. J.F. Mas, R. González, Change detection and land use/land cover database updating using image segmentation. in *GIS Analysis and Visual Interpretation. ISPRS Geospatial Week, 28 Sep–03 Oct, La Grande Motte, France*, 2015
7. M. Castelluccio, G. Poggi, C. Sansone, L. Verdoliva, Land use classification in remote sensing images by convolutional neural networks. arXiv preprint arXiv:1508. 00092, 2015
8. B. Zhao, B. Huang, Y. Zhong, Transfer learning with fully pretrained deep convolution networks for land-use classification. IEEE Geosci. Remote Sens. Lett. **14**(9), 1436–1440 (2017)

9. S. Jiang, A. Alves, F. Rodrigues, et al., Mining point-of-interest data from social networks for urban land use classification and disaggregation. Comput. Environ. Urban Syst. **53**, 36–46 (2015)

10. S. Paldino, I. Bojic, S. Sobolevsky, et al., Urban magnetism through the lens of geo-tagged photography. EPJ Data Sci. **4**(1), 5 (2015)

11. X. Deng, S. Newsam, Quantitative comparison of open-source data for fine-grain mapping of land use. in *ACM SIGSPATIAL International Conference on Advances in Geographic Information Systems*, 2017

12. L. Wang, F. Fang, X. Yuan, et al., Urban function zoning using geotagged photos and openstreetmap. Geosci. Remote Sens. Symp. **2017**, 815–818 (2017)

13. L. Cheng, S.S. Chu, W.W. Zong, et al., Use of tencent street view imagery for visual perception of streets. ISPRS Int. J. Geo-Inf. **6**(9), 265 (2017)

14. J. Kang, M. Körner, Y. Wang, H. Taubenböck, X.X. Zhu, Building instance classification using street view images, ISPRS J Photogramm. Remote Sens. 145(Part A), 44–59 (2018)

15. I. Seiferling, N. Naik, C. Ratti, et al., Green streets − quantifying and mapping urban trees with street-level imagery and computer vision. Landsc. Urban Plan. **165**, 93–101 (2017)

16. L. Liu, E.A. Silva, C. Wu, et al., A machine learning-based method for the large-scale evaluation of the qualities of the urban environment. Comput. Environ. Urban Syst. **65**, 113–125 (2017)

17. X. Li, C. Zhang, W. Li, Building block level urban land-use information retrieval based on Google Street View images. Gisci. Remote Sens. **2017**, 1–17 (2017)

18. S. Branson, J.D. Wegner, D. Hall, et al., From Google Maps to a fine-grained catalog of street trees. ISPRS J. Photogramm. Remote Sens. **135**, 13–30 (2018)

19. S. Karayev, M. Trentacoste, H. Han, et al., Recognizing image style. arXiv preprint arXiv:1311.3715, 2013

20. C. Szegedy, S. Ioffe, V. Vanhoucke, et al., Inception-v4, inception-ResNet and the impact of residual connections on learning. Comput. Vision Pattern Recogn. 2016

21. O. Russakovsky, J. Deng, H. Su, et al., ImageNet large scale visual recognition challenge. Int. J. Comput. Vis. **115**, 3 (2014)

22. H. Rao, X. Shi, A. K. Rodrigue, J. Feng, Y. Xia, M. Elhoseny, X. Yuan, L. Gu, Feature selection based on artificial bee colony and gradient boosting decision tree Appl. Soft Comput. 2018. https://doi.org/10.1016/j.asoc.2018.10.036

23. M. Elhoseny, K. Shankar, J. Uthayakumar, Intelligent diagnostic prediction and classification system for chronic kidney disease. Nat. Sci. Rep. 2019. https://doi.org/10.1038/s41598-019-46074-2

24. N. Krishnaraj, M. Elhoseny, M. Thenmozhi, Mahmoud M. Selim, K. Shankar, Deep learning model for real-time image compression in Internet of Underwater Things (IoUT). J. Real-Time Image Process. 2019. https://doi.org/10.1007/s11554-019-00879-6

25. X. Yuan, V. Sarma, Automatic urban water-body detection and segmentation from sparse ALSM data via spatially constrained model-driven clustering. IEEE Geosci. Remote Sens. Lett. **8**(1), 73–77 (2010)

26. B.S. Murugan, M. Elhoseny, K. Shankar, J. Uthayakumar, Region-based scalable smart system for anomaly detection in pedestrian walkways. Comput. Electr. Eng. **75**, 146–160 (2019)

Chinese–Vietnamese Bilingual News Event Summarization Based on Distributed Graph Ranking

Shengxiang Gao, Zhengtao Yu, Yunlong Li, Yusen Wang, and Yafei Zhang

Abstract Multi-language news event summarization aims to quickly obtain important information from lots of related news texts written in different languages automatically. Considering that the main expressed information for the same event is similar no matter what language it is presented, the paper proposes a novel unified approach to summarize important information from the monolingual and Chinese–Vietnamese bilingual news sets simultaneously. Firstly, analyzing the sentence dependence relationship, making rules to segment sentences into different grammatical parts, a bilingual dictionary is used to set up a bilingual feature space. Secondly, Chinese–Vietnamese sentence graph model is calculated distributively. Finally, using the features that graph nodes can boost each other and fusing context information, the sentences are ranked based on whether they can represent the important information. The experimental result shows that our method is effective.

Keywords Event summarization · Distributed graph ranking · Multi-language multi-document · Chinese–Vietnamese

1 Introduction

With the rapid development of the Internet, online news has exploded every day, and online news text analysis is becoming more and more important for web content security detection and rapid access to network information. Text summarization is an effective way to analyze web text. It can be modeled by machine learning, which extracts important sentences to capture the central ideas of web text. These sentences help us to quickly understand a news text. The same topic news is often described by

S. Gao · Z. Yu (✉) · Y. Li · Y. Wang · Y. Zhang
Faculty of Information Engineering and Automation, Kunming University of Science and Technology, Kunming, China

Yunnan Key Laboratory of Artificial Intelligence, Kunming University of Science and Technology, Kunming, China

© Springer Nature Switzerland AG 2020
X. Yuan, M. Elhoseny (eds.), *Urban Intelligence and Applications*, Studies in Distributed Intelligence, https://doi.org/10.1007/978-3-030-45099-1_8

different languages. Massive multilingual news texts in different languages make it hard to machine understanding. Therefore, it is necessary to summarize multilingual news text sets automatically and generate concise text which can help people get the main information of a multilingual event quickly.

Nowadays, text summarization [1] techniques can be divided into two categories: abstractive [2] or extractive [3] summarization. An extractive summarization is generated by selecting a few relevant sentences from an original document. Whereas abstractive summarization produces an abstract summary which includes words and phrases different from the ones occurring in the source document. It needs extensive natural language processing or large-scale corpus. At present, there are relatively few bilingual corpus resources in Chinese and Vietnamese, and the language differences between Chinese and Vietnamese are large. Abstract summarization is difficult to be realized. Therefore, this paper focuses on extracting important sentences from Chinese and Vietnamese news articles.

In the existing extractive summarization methods, multi-document summarization based on the graph ranking algorithm [4] achieves good results. Its main idea is to take sentences as nodes in a graph, and each edge in the graph is the relation between two sentences. Then, based on the established graph model, every sentence is given a score by ranking algorithm. Finally, sentences with high scores will be selected as a summary. The key problem of this method is how to establish the relationship between sentences and how to effectively integrate the relationship into the ranking algorithm. When addressing Chinese–Vietnamese news event articles, with two language expressions and different grammatical order, it is very hard to establish relationships between Chinese sentences and Vietnamese sentences. For example, in Vietnamese, adjectives usually follow nouns, but Chinese is exactly the opposite. In addition, the position of adverbial ingredients in Chinese is relatively flexible; however, in Vietnamese, adverbials are usually placed in front. Examples are as follows:

The meaning of sentences written in Chinese and Vietnamese in Fig. 1 is consistent, and the meaning is "my black cow grazes on the fields in the village". The meaning of the word pairs at both ends of the arrow is consistent. So, the grammatical difference between the two languages is large.

At present, many scholars use machine translation technology to achieve cross-language understanding, but Chinese–Vietnamese machine translation has not achieved satisfactory results, which brings difficulties to our work.

In this paper, for sentence similarity, considering the grammatical differences between Chinese and Vietnamese, sentences are divided into different grammatical components according to rules. And the similarity between sentences is calculated based on the similarity of different grammatical components. For ranking process-

Fig. 1 Sentence comparison between Chinese and Vietnamese

ing, consider that if a sentence can cover the main information of monolingual news articles, to some extent, it also can cover the main information of bilingual news article set. We propose a novel method to summarize monolingual document and bilingual document simultaneously by integrating the mutual influences between the two tasks into ranking algorithms. The main constructions are summarized as follows:

1. Mapping Chinese and Vietnamese news texts into an undirected graph, its nodes represent sentences (in Chinese or Vietnamese) and its edges represent the similarity between two sentences (Chinese to Chinese, Chinese to Vietnamese, or Vietnamese to Vietnamese);
2. Analyzing the sentence dependence relationship, making rules to segment a sentence into different grammatical parts;
3. The similarity between two sentences is computed by a weighted average of the similarities between their subject compositions, predicate compositions, object compositions, and other compositions. If it is the similarity between a Chinese sentence and a Vietnamese sentence, on the basis of bridged bilingual feature space, the same way works well;
4. Distributed graph ranking is employed to decrease the time cost for solving the model. And it effectively extracts the important sentences from the Chinese and Vietnamese news texts to form their summary.

The rest of this paper is organized as follows: Sect. 2 reviews related work. Our proposed method is presented in Sect. 3. Section 4 shows the experiments and results. Lastly, we conclude this paper in Sect. 5.

2 Related work

Extractive summary, according to the number of documents, language characteristics, can be divided into three types: single-document summary, multi-document summary, multi-language multi-document summary. The purpose of single-document summarization is to extract some sentences which can describe the important information from a news text. This method mainly considers context statistical features. At first, many scholars have evaluated the importance of a sentence by combining word frequency, sentence position, keyword, sentence length, clue word, etc. [5]. With the rise of machine learning, some scholars extract important sentences by machine learning methods [6, 7].

The purpose of multi-document summarization is to extract some sentences that describe their important information from a large number of news texts that describe the same event. In recent years, the method of multi-document summarization based on graph sorting has achieved good results. This method uses the sentence as the node of the graph. The association between sentences is represented as a document, and the document is represented as a document map. Then, the weight of nodes is calculated based on the graph ranking algorithm. Then, based on the results of graph sorting, the summarization and smoothing strategies are adopted to select the abstract sentences, so as to generate event summaries. Wan [8] makes use of the interaction between sentences and documents in a text to get important sentences in the collection of event documents. Baralis et al. [4] build a graph model by mining association rules of the elements in the document, thus obtaining important sentences. Erkan et al. [9] use LexRank algorithm to solve the graph model and have achieved good results. Later scholars put topic models into graph models to extract the abstract sentence. This method takes account of the semantic correlation behind the text, calculates the correlation between the topic and the sentence, and gets the event summary. Li et al. [10] extract important sentences under important topics through the probability distribution of sentences on the topic. Xu et al. [11] build relationships between topics and integrate them into an event summary method by refining the existing topic model. Since graph computing requires high time complexity, in order to improve graph computing effectiveness, there are studies about parallel or distributed graph computation that have made good results [13–16, 20].

Multilingual and multi-document summarization is currently less researched, mainly based on machine translation's method. The effect of these methods is limited by Machine Translation's effect. Most of the researches are committed to reducing the error brought by Machine Translation. Wan et al. [12] first rating of each sentence, and then scoring according to the degree of importance of the sentence in the text, two comprehensive scores of sentences, will be able to express the original meaning. After they are translated, the target sentences are regarded as the summarization of the text.

3   Our Proposed Methodology

3.1   Chinese–Vietnamese News Event Graph Model

As illustrated in Fig. 2, given a set of Chinese and Vietnamese news articles describing the same event, we map these bilingual news texts into an undirected graph. In this undirected graph, the nodes represent sentences written in Chinese or Vietnamese, and the edges represent the similarity between two sentences.

The nodes on the left part of Fig. 2 represent Chinese sentences, and the nodes on the right part of Fig. 2 represent Vietnamese sentences. In Fig. 2, there are four types of edges, formally as E^{cn} represents the similarity between Chinese sentences. E^{ve} represents the similarity between Vietnamese sentences. E^{vecn}, E^{cnve} represent the similarity between Chinese and Vietnamese sentences.

When calculating sentence similarity, the traditional method is realized through word co-occurrence. That is, two sentences are related if they contain the same words. This method does not consider the influence of different grammatical

Fig. 2 Chinese–Vietnamese news event graph model

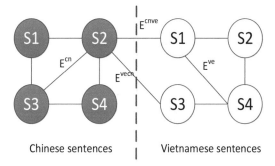

components in sentences on sentence similarity. For example, the following two sentences:

> Nguyen Phu Chong invited Xi Jinping to pay a state visit;
>   Xi Jinping invited Ngurah Phuong Fu to pay a state visit;

If we use the method of words co-occurrence to measure the similarity of the above sentences, the meanings of the two sentences are the same. In fact, as the subject and predicate reversed, their meaning is different, and the similarity value should be small. To solve this problem, the feature of word order is considered to calculate sentence similarity [17]. In the monolingual environment, due to the relatively fixed grammar, this method achieved good results. However, for the similarity calculation of Chinese and Vietnamese bilingual sentences, the Chinese and Vietnamese grammars are greatly different. Therefore, it is difficult to apply directly them, as shown in the example sentences in the introduction section.

Through reading and analyzing the bilingual news sentence written in Chinese and Vietnamese, it is found that for each news sentence, the main information it contains can be characterized as "someone does something," corresponding to the grammatical representation of "subject, predicate, object." Therefore, to avoid this problem caused by the differences in grammar between Chinese and Vietnamese, we segment bilingual sentences into different grammatical components, and then the similarity between two sentences is calculated based on the similarity of different grammatical components. Accordingly, we map the different grammatical components of bilingual sentences into the same feature space, and each sentence can be characterized as follows:

$$\Big\{\{\text{subject}\}, \Big\{\text{verb}\Big\}, \{\text{object}\}, \{\text{other}\}\Big\},$$

where {subject}, {verb}, {object}, and {other} represent the subject, predicate, object, other components in the sentence, respectively. Then the similarity is obtained by calculating the similarities of different grammatical components separately, specifically, divided into two steps:

3.1.1 Vietnamese Sentence Syntax Component Extraction

Given a sentence, we make rules to obtain its subject, predicate, object composition based on sentence dependencies. For Chinese sentences, we use Stanford Parser to get dependencies between words. For Vietnamese sentences, VSDP Parser is used.

The rules to discriminate subject, predicate, object composition are as follows:

1. Predicate composition: The word "root" refers to and the words that have a side-by-side relationship ("conj" labeled) with the word (we call these words "predicate words");
2. Subject composition: The words that have a subject-predicate relationship ("nsubj" labeled) with the "predicate words" and all words in the subtree with those words as the root node (we call these words "subject words");
3. Object composition: The words that has a predicate-object relationship ("dobj" labeled) with the "predicate words" and the words that has a side-by-side relationship ("conj" labeled) with those words and all words in the subtree with above words as the root node (we call these words "object words");
4. Other composition: Other words in the sentence. (We call these words "other words").

According to the above rules, a sentence can be converted into the following form:

$$S = \{\{s_1, \ldots, s_i, \ldots, s_a\}, \{v_1, \ldots, v_i, \ldots, v_b\},$$
$$\{ob_1, \ldots, ob_i, \ldots, ob_c\}, \{ot_1, \ldots, ot_i, \ldots, ot_e\}\},$$

where s_i, v_i, ob_i, ot_i represent "subject words," "predicate words," "object words," and "other words," respectively. For each part, considering that news elements have a better ability to express news sentences, we only keep verbs, nouns, time words, place words, person nouns, and organization names. Due to the lack of a tool for extracting news elements from Vietnamese news, this paper simply completes the character matching of Vietnamese words with pre-collected Vietnamese entities.

3.1.2 Chinese–Vietnamese Sentence Similarity Calculation

Chinese–Vietnamese sentence similarity calculation: for two sentences in the same language, firstly, we calculate the subject composition similarity. Suppose the subject composition of sentences S_1 and S_2 is as follows:

$$S_1 = \{s_1, \ldots, s_e\}$$
$$S_2 = \{s_1, \ldots, s_g\}$$

Each dimension represents the words in the subject composition. We unite the words of two sentences and get subject composition vector feature space

$S_{1-2} = \{s_1, \ldots, s_g, s_h, \ldots, s_m\}$, $m \le e + g$. Then we map the two sentences' subject composition into the new feature space, if the sentence contains a word of a certain dimension, the value of the dimension is 1, else the value is 0. Then the cosine similarity is used to calculate the similarity of the subject composition. Thus, we obtain the subject composition similarity, formalized as sim_s. The similarities of predicate composition, object composition, other composition are calculated in the same way, formalized as sim_v, sim_o, sim_{ot}.

Finally, the similarity of two sentences in the same language is calculated from the following formula:

$$\text{sim} = a * \text{sim}_s + a * \text{sim}_v + a * \text{sim}_o + b * \text{sim}_{ot} \tag{1}$$

According to the analysis, the subject composition, the predicate composition, the object composition contribute more than other components. Therefore, we set $a = 0.2$, $b = 0.22$. For Chinese and Vietnamese sentence similarity calculation, Chinese–Vietnamese dictionary is used to build a bilingual feature space for each syntax component:

$$S_{cv} = \{(e_{c1}, e_{v1}), (e_{c2}, e_{v2}), \ldots, (e_{ci}, e_{vi})\}$$

Each dimension represents a Chinese–Vietnamese mutual translation pair with the same meaning. Then we calculate the similarity of two bilingual sentences in the same way.

Finally, we can obtain a similarity matrix between Chinese sentences, formalized as $M_{\text{ch\_ch}}$, similarity matrix between Vietnamese sentences, formalized as $M_{\text{vie\_vie}}$, similarity matrix between Chinese sentences and Vietnamese sentences, formalized as $M_{\text{ch\_vie}}$, $M_{\text{vie\_ch}}$, and $M_{\text{chvie}}^T = M_{\text{viech}}$.

3.2 A Solution to Chinese–Vietnamese News Event Graph Model

We propose a novel ranking algorithm for Chinese–Vietnamese news event summarization. In the processing of ranking algorithm, each sentence is given two scores, representing abilities to describe the main information of monolingual document set and bilingual document set, respectively. This idea is based on the assumption that if a sentence can cover the main information of monolingual news articles, to some extent, it also can cover the main information of the bilingual news article set. It is reasonable because these articles describe the same news event and their news elements are similar. Specifically, CR1 indicates the ability of which Chinese sentences represent the main information of a Chinese document set; CR2 indicates the ability of which Chinese sentences represent the main information of Chinese–Vietnamese document set; VR1 indicates the ability of which Vietnamese sentences represent the main information of a Vietnamese document set; VR2 indicates the ability of which Vietnamese sentences represent the main information of Chinese–Vietnamese document set. Their value is calculated based on the following assumptions:

1. If a Chinese sentence is associated with other sentences with a higher CR1 score in the Chinese document set, then the sentence has a higher CR1 score;
2. If a Chinese sentence is associated with a sentence with a higher VR2 score in the Vietnamese document collection, then the sentence has a higher CR1 score;
3. If a Chinese sentence is related to other sentences with a higher CR1 score in the Chinese document set, then the sentence has a higher CR2 score;
4. If a Chinese sentence is associated with a sentence with a higher VR2 score in the Vietnamese document set, then the sentence has a higher CR2 score;
5. If a Chinese sentence is related to other sentences with a higher CR2 score in the Chinese document set, then the sentence has a higher CR2 score.

Base on the above assumptions, we get sentence score calculation formula:

$$CR1 = \alpha_1 M_{chch}CR1 + \beta_1 M_{chvie}VR2 \tag{2}$$

$$VR1 = \beta_1 M_{vievie}VR1 + \alpha_1 M_{viech}CR2 \tag{3}$$

$$CR2 = \alpha_2 M_{chch}CR1 + \beta_2 M_{chch}CR2 + \gamma_2 M_{chvie}VR2 \tag{4}$$

$$VR2 = \alpha_2 M_{vievie}CR2 + \gamma_2 M_{viech}CR2 + \beta_2 M_{vievie}VR2 \tag{5}$$

When calculating the ability of a sentence to represent the main information of a monolingual document set, α_1 represents the sentence in the document represent the weight of the document, β_1 represents the sentences in other documents represent the weight of the document when calculating the ability of a sentence to represent the main information of a bilingual document set, α_2 represents the sentence in the document represent the weight of the document, β_2 represents the sentences in other documents represent the weight of the document, γ_2 represents the sentences in other documents represent the weight of the bilingual document.

However, the above work considers the association of sentences with other sentences within the whole text collection, ignoring the contextual relationships. According to the characteristics of news reports, the title of a report tends to reflect its important information and has important indications for the production of summary. The more similar the sentence is to the title, the more likely it is to represent the main information of this article, the more likely it is to represent the documentation set's main information. Here, we use $P1$ to represent the similarity between the Chinese title and the sentence in the Chinese document, $P2$ to indicate the similarity between the Vietnamese title and the sentence in the Vietnamese document, $P3$ to indicate the similarity between the Chinese title and the sentence in the Chinese and Vietnamese bilingual documents, $P4$ indicates the similarity between the Vietnamese title and the sentence in the Chinese and Vietnamese bilingual documents. Further, the formula is as follows:

$$CR1 = \alpha_1 M_{chch} CR1 + \beta_1 M_{chvie} VR2 + \xi P1 \qquad (6)$$

$$VR1 = \beta_1 M_{vievie} VR1 + \alpha_1 M_{viech} CR2 + \xi P2 \qquad (7)$$

$$CR2 = \alpha_2 M_{chch} CR1 + \beta_2 M_{chch} CR2 + \gamma_2 M_{chvie} VR2 + \xi P3 \qquad (8)$$

$$VR2 = \alpha_2 M_{vievie} CR2 + \gamma_2 M_{viech} CR2 + \beta_2 M_{vievie} VR2 + \xi P4 \qquad (9)$$

Among them, ξ indicates the weight of the similarity between the sentence and the title of the document.

Each dimension of vector $P1$, $P2$, $P3$, and $P4$ represents the similarity between the sentence and the title of the document in which the sentence is located. For example, the formula for $P1$ is as follows:

$$P1[i] = 0.5 + \frac{S_i S_{title}}{2 \times |S_i| |S_{title}|} \qquad (10)$$

where S_i represents the Chinese sentence, S_{title} is the title of the Chinese document where the sentence is located. $P2$, $P3$, $P4$ is defined in the same way. Further, we define

$$M = \begin{bmatrix} \alpha_1 M_{chch} & 0 & 0 & \beta_1 M_{chvie} \\ 0 & \beta_1 M_{vievie} & \alpha_1 M_{viech} & 0 \\ \alpha_2 M_{chch} & 0 & \beta_2 M_{chch} & \gamma_2 M_{chvie} \\ 0 & \alpha_2 M_{vievie} & \gamma_2 M_{viech} & \beta_2 M_{vievie} \end{bmatrix}, R = \begin{bmatrix} CR1 \\ VR1 \\ CR2 \\ VR2 \end{bmatrix}, P = \begin{bmatrix} \xi P1 \\ \xi P2 \\ \xi P3 \\ \xi P4 \end{bmatrix}$$

Then the target formula is transformed as follows:

$$(I - M) R = P \qquad (11)$$

According to the above deduction, the task of Chinese–Vietnamese bilingual news events summarization is transformed into solving the problem of linear equations. The solution is the sentence score. To guarantee the solution of the linear system Eq. (11), we make the following two transformations on M. First M is normalized by columns. If all the elements in a column are zero, we replace zero elements with $1/n$ (n is the total number of the elements in that column). Second, M is multiplied by a decay factor ($0 < \theta < 1$), such that each element in M is scaled-down but the meaning of M will not be changed. Finally, Eq. (11) is rewritten as:

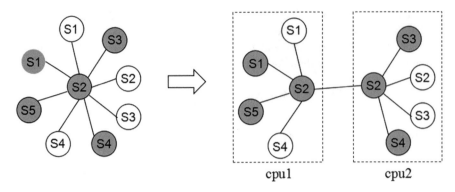

Fig. 3 Greedy point segmentation strategy

$$(I - \theta M) R = P \qquad (12)$$

The matrix $I - \theta M$ is a strictly diagonally dominant matrix now, and the solution of the linear system Eq. (12) exists. Now, we are ready to adopt the Gauss–Seidel method to solve the linear system Eq. (12).

Because there is a high time complexity for graph ranking, this paper uses distributed computing to optimize the graph model construction and reduce the time complexity of graph sorting. The PowerGraph graph processing system is used to traverse the graph with the vertex as the granularity of the calculation. The method divides the nodes in the graph by the greedy strategy, so that each edge of the text in the graph appears in a subgraph. The specific segmentation strategy is shown in Fig. 3. The method implements information interaction between graph nodes based on shared memory and constructs graph ranking model through asynchronous scheduling.

3.3 Summary Generation

According to the systematical solution to linear equations, we can get the scores of each sentence. The higher the score, the more likely it is to be selected as a summary. However, there are a large number of sentences with similar meanings tend to have similar scores in the ranking process. Therefore, we need to select sentences with high scores and different meanings. This paper adopts a greedy algorithm to generate Summary of Chinese Documents, Summary of Vietnamese Documents, Summary of Chinese–Vietnamese bilingual documents, respectively. For a sentence set $S = \{s_1, s_2, \ldots, s_n\}$ and a sentence score function $f(s_i)$, the algorithm details are as follows:

Algorithm: summary generation

1: Initialization: $Y = \Phi$, $X = \{S - s_1\}$;
2: While the number of sentences in Y is less than 5;
3:     $s_m = \mathrm{argmax}_{s_i \in X} f(s_i)$;
4:     **if** $\mathrm{sim}(s_m, s) < \mathrm{Th}_{\mathrm{sem}}$, **for** all $s \in Y$:
5:         $Y = Y + \{s_m\}$;
6:     $X = x - \{s_m\}$;
7: **END**

Here, Y is the output summary. s a sentence in Y. X is the input sentence set. s_i represents a sentence in S. s_m is the sentence in X which makes $f(s_i)$ to reach the maximum. $\mathrm{Th}_{\mathrm{sem}}$ is the Similarity threshold.

4 Experimental Studies

4.1 Datasets and Evaluation Metrics

Up to now, no relevant benchmark dataset has been found for Chinese–Vietnamese bilingual news event summarization. In order to evaluate the proposed method, four hot news events are selected. Then a few news articles about each topic, in the Chinese and Vietnamese languages, are collected. For each topic, we selected the articles in the two languages within the same period. They are published by major Chinese and Western news agencies. The statistics of the corpus are summarized in Table 1.

For each news topic, we read these articles carefully and select five sentences from Chinese articles as a Chinese summary, five sentences from Vietnamese articles as Vietnamese summary, five sentences from Chinese articles and Vietnamese as bilingual summary.

Table 1 Chinese and Vietnamese news events corpus

Topic	Language	The number of texts
Nguyen Phu Trong visits China	Chinese	55
	Vietnamese	50
Mekong drought	Chinese	45
	Vietnamese	38
Defense Minister meeting	Chinese	62
	Vietnamese	35
Food delivery market	Chinese	74
	Vietnamese	45

We adopt the widely used ROUGE toolkit[1] to evaluate our model and report recall results on three metrics of ROUGE-1, ROUGE-2, and ROUGE-SU4 [18].

4.2 Baselines

For comparison, we select the following three models as baselines including:

1. Mt-UnifiedRank: Firstly, we translate the Vietnamese news articles into Chinese by means of the Baidu translation tool. Then we also treat the Chinese and Vietnamese documents as two sets and use the method proposed in this paper to construct the Chinese–Vietnamese bilingual news sentence graph model. Finally, we generate a summary for Chinese articles, Vietnamese articles, and Chinese–Vietnamese bilingual articles, separately;
2. Sim-UnifiedRank: Firstly, with the help of Chinese–Vietnamese dictionaries, sentences similarity is calculated by words co-occurrence, then we use the ranking algorithm proposed in this paper to generate summary;
3. IterRank: Firstly, with the help of Chinese–Vietnamese dictionaries, a bilingual feature space is constructed. The sentence similarity is then calculated by the method we proposed and a bilingual news sentence graph model is constructed, and an improved PageRank algorithm [19] is used to generate a summary for Chinese–Vietnamese bilingual articles.

To further verify the effectiveness of the proposed method, we also apply our model to generate monolingual document summarization in Chinese and Vietnamese, respectively. For comparison, we select the Mono-IterRank model proposed in [19] as the baseline.

4.3 Experimental Settings

In this paper, we tune our model in the dataset to determine the optimal parameters. Firstly, when calculating the ability of a sentence to represent the main information of a monolingual document set, we select the weight of that a sentence in a document represents the document as $\alpha_1 = 0.3$ and the weight of those sentences in other documents represent the document as $\beta_1 = 0.6$. Secondly, when calculating the ability of a sentence to represent the main information of a bilingual document set, we select the weight of that a sentence in a document represents the document as $\alpha_2 = 0.1$, the weight of those sentences in other documents represent the document as $\beta_2 = 0.5$ and the weight of those sentences in other documents represent the bilingual documents as $\gamma_2 = 0.3$. Finally, we select the weight of the similarity

---

[1] https://github.com/bheinzerling/pyrouge.

between the sentence and the title of the document as $\xi = 0.1$. Th_{sem}, the similarity threshold, is set to 0.5 according to past experiences.

4.4   Results

For bilingual-document summarization, the results of our method and the baseline methods are summarized in Table 2. Our method achieves a ROUGE-1 score of 0.3412, a ROUGE-2 score of 0.2332, and a ROUGE-SU4 score of 0.2912, outperforming all baseline methods.

Firstly, we compare the results of the proposed method and Mt-UnifiedRank, we observe that the proposed method improved by 0.076 on ROUGE-SU4 than Mt-UnifiedRank, suggesting that using Chinese–Vietnamese bilingual dictionary to construct the Chinese–Vietnamese bilingual feature spaces get a better effect than machine translation because of the low accuracy of Chinese–Vietnamese machine translation. Secondly, we compare the results of the proposed method and Sim-UnifiedRank, we observe that the proposed method improved by 0.055 on ROUGE-SU4 than Sim-UnifiedRank, we can conclude that segmenting the sentence into different grammatical components, then calculating sentence similarity is more accurate because word order has an important effect on sentences similarity. Finally, we compare the results of the proposed method and Mt-UnifiedRank, and observe that the proposed method improves by 0.033 on ROUGE-SU4 than IterRank. This suggests that our ranking algorithm is more effective.

We also test our method on the monolingual summary task. The results are shown in Table 3. The results indicate that our approach outperforms the baseline method over all metrics in both languages.

We compare the performance of our method with Mono-IterRank, we see 0.085 ROUGE-1 improvements in Chinese and 0.062 ROUGE-1 improvements in Vietnamese. On ROUGE-2, our method makes an improvement of 0.134 in Chinese and 0.021 in Vietnamese. On ROUGE-SU4, our method makes an improvement

Table 2 Comparison results of different bilingual summarization methods

System	ROUGE-1	ROUGE-2	ROUGE-SU4
Mt-UnifiedRank	0.2826	0.1893	0.2152
Sim-UnifiedRank	0.3011	0.2098	0.2359
IterRank	0.2996	0.2131	0.2587
Ours	0.3412	0.2332	0.2912

Table 3 Comparison Results of different monolingual summarization methods

System	ROUGE-1		ROUGE-2		ROUGE-SU4	
	Chinese	Vietnamese	Chinese	Vietnamese	Chinese	Vietnamese
Mono-IterRank	0.3132	0.2519	0.2011	0.2641	0.3287	0.2688
Ours	0.3982	0.3142	0.3352	0.2851	0.3821	0.3091

Table 4 Comparison of time
efficiency of different
methods

System	Time (s)
IterRank	325
Ours-without-dis	430
Ours-with-dis	86

of 0.053 in Chinese and 0.04 in Vietnamese. The experiment proves that the monolingual document and bilingual document summarizations can benefit each other by making use of mutual influences.

4.5   Time Efficiency Analysis

To verify time efficiency, we compare the calculating time of the proposed method with or without distributed computing on our dataset. The experiment is conducted on a single server and a cluster of three workers individually. Each server consisted of Intel(R) Core(TM) i7-7700, 3.6 GHz CPU, 8 GB RAM, 1 TB hard drive, and runs Windows 10 operating system. The results are shown in Table 4. Here, the proposed method is run without distributed computing is marked as Ours-without-dis. That the proposed method is run with distributed computing is marked as Our-with-dis.

From Table 4, on the dataset, our-without-dis has 430 s of runtime, and Ours-with-dis only spends 86 s. After added distributing computation during graph ranking, the proposed method is faster largely and saves the runtime to 20%. Compared to IterRank, Ours-with-dis is four times faster than it. It is obvious that distributed computing can greatly improve the solving speed of the graph model.

5   Conclusion and Future Work

For Chinese and Vietnamese, bilingual word co-occurrence, bilingual differences in grammar and word order, and a bilingual dictionary, which can be used to construct a bilingual feature space, will effectively supervise similarity calculation among monolingual sentences or bilingual sentences. Therefore, we propose a novel unified approach to simultaneously generate monolingual document and bilingual document summarization by making using of the mutual influences between the two tasks. Experimental results show that the proposed approach is effective. In future work, combining bilingual language differences and characteristics of multi-language texts, it may be worth discussing for generated summarization of multi-language web news texts based on deep learning.

Acknowledgments The work was supported by National Natural Science Foundation of China (Grant Nos. 61972186, 61732005, 61761026, 61672271 and 61762056), National Key Research and Development Plan (Grant Nos. 2018YFC0830105, 2018YFC0830100), Yunnan high-tech industry development project (Grant No. 201606), Natural Science Foundation of Yunnan Province (Grant No. 2018FB104), and Talent Fund for Kunming University of Science and Technology (Grant No. KKSY201703005).

References

1. M. Gambhir, V. Gupta, Recent automatic text summarization techniques: a survey. Artif. Intell. Rev. **47**(1), 1–66 (2017)
2. S. Chopra, M. Auli, A.M. Rush, Abstractive sentence summarization with attentive recurrent neural networks, in *Proceedings of 2016 Conference of the North American Chapter of the Association for Computational Linguistics: Human Language Technologies, NAACL HLT 2016* (ACL, San Diego, 2016), pp. 93–98
3. K. Hong, J.M. Conroy, B. Favre, A. Kuesza, H. Lin, A. Nenkova, A repository of state of the art and competitive baseline summaries for generic news summarization, in *Proceedings of the 9th International Conference on Language Resources and Evaluation, LREC 2014* (ELRA, Reykjavik, 2014), pp. 1608–1616
4. E. Baralis, L. Cagliero, N. Mahoto, A. Fiori, GRAPHSUM: discovering correlations among multiple terms for graph-based summarization. Inf. Sci. **249**, 96–109 (2013)
5. H.P. Luhn, The automatic creation of literature abstracts. IBM J. Res. Dev. **2**(2), 159–165 (1958)
6. D. Shen, J.T. Sun, H. Li, Q. Yang, Z. Chen, Document summarization using conditional random fields, in *Proceedings of 20th International Joint Conference on Artificial Intelligence, IJCAI 2007* (IJCAI, Morgan Kaufmann, Hyderabad, 2007), pp. 2862–2867
7. L. Li, K. Zhou, G.R. Xue, H. Zha, Y. Yu, Enhancing diversity, coverage and balance for summarization through structure learning, in *Proceedings of the 18th International World Wide Web Conference, WWW 2009* (ACM, Madrid, 2009), pp. 71–80
8. X. Wan, Towards a unified approach to simultaneous single-document and multi-document summarizations, in *Proceedings of the 23rd International Conference on Computational Linguistics, Coling 2010* (ACM, Beijing, 2010), pp. 1137–1145
9. G. Erkan, D.R. Radev, LexRank: graph-based lexical centrality as salience in text summarization. J. Artif. Intell. Res. **22**(204), 457–479 (2004)
10. Y. Li, S. Li, Query-focused multi-document summarization: combining a novel topic model with graph-based semi-supervised learning, in *Proceedings of the International Conference on Computational Linguistics, Coling 2014* (ACM, Dublin, 2014), pp. 1197–1207
11. J.A. Xu, J.M. Liu, K. Araki, A hybrid topic model for multi-document summarization. IEICE Trans. Inf. Syst. **98**(5), 1089–1094 (2014)
12. X. Wan, H. Li, J. Xiao, Cross-language document summarization based on machine translation quality prediction, in *Proceeding of the Annual Meeting of the Association for Computational Linguistics, ACL2010* (ACL, Uppsala, 2010), pp. 917–926
13. J.F. García, M.V. Carriegos, On parallel computation of centrality measures of graphs. J. Supercomput. **75**(3), 1410–1428 (2019)
14. M. Nasir, K. Muhammad, J. Lloret, A.K. Sangaiah, M. Sajjad, Fog computing enabled cost-effective distributed summarization of surveillance videos for smart cities. J. Parallel Distrib. Comput. **126**, 161–170 (2019)
15. J. Samuel, X. Yuan, X. Yuan, B. Walton, Mining online full-text literature for novel protein interaction discovery. in *Proceeding of International Workshop on Data Mining for High Throughput data from Genome-Wide Association Studies. IEEE Int'l Conf. on Bioinformatics & Biomedicine, Hong Kong, Dec 18–21*, 2010
16. L. Gu, Y. Han, C. Wang, W. Chen, J. Jiao, X. Yuan, Module overlapping structure detection in PPI using an improved link similarity-based Markov clustering algorithm. Neural Comput. & Applic. **31**(5), 1481–1490 (2018)

17. Y. Li, D. McLean, Z.A. Bandar, J.D. O'Shea, K. Crockett, Sentence similarity based on semantic nets and corpus statistics. IEEE Trans. Knowl. Data Eng. **18**(8), 1138–1150 (2006)
18. C.Y. Lin, E. Hovy, Automatic evaluation of summaries using n-gram co-occurrence statistics, in *Proceedings of the 2003 Human Language Technology Conference of the North American Chapter of the Association for Computational Linguistics, NAACL 2003* (NAACL, Edmonton, 2003), pp. 150–157
19. R. Mihalcea, P. Tarau, A language independent algorithm for single and multiple document summarization. Unt Sch. Works **2005**, 19–24 (2005)
20. X. Yuan, J. Zhang, X. Yuan, B.P. Buckles, Multi-scale feature identification using evolution strategies. Image Vis. Comput. **23**(6), 555–563 (2005)

Entity Hyponymy Extraction of Complex Sentence Combining Bootstrapping and At-BiLSTM in Special Domain

Huaqin Li, Zhiju Zhang, Zhengtao Yu, Hongbin Wang, and Hua Lai

Abstract Acquiring entity hyponymy of complex sentences can be a highly difficult process in special domain. To tackle this problem, this paper proposes a novel
method that combines Bootstrapping method and Attention-Based Bidirectional
Long Short-Term Memory Networks (Bo-At-BiLSTM). The experimental corpus is
in the field of tourism in China. First, the bootstrapping method is used to obtain the
patterns set. Then, pattern matching is employed to acquire the candidate sentences
and word embedding. Next, import into the bidirectional Long Short-Term Memory
Networks and introduce attention mechanism. Finally, output the results by Softmax
classifier. The experimental results on the tourism corpus show that the proposed
approach outperforms the baseline methods.

Keywords Hyponymy extraction · Complex sentences · Bootstrapping method ·
At-BiLSTM

1 Introduction

The hyponymy, one of the important semantic relations, is the basis for describing
the hierarchical relationship of things [1]. The extraction of entity hyponymy is
an indispensable part of building the domain knowledge graphs. Currently, the
definition of hyponymy given by Miller [2] is adopted, "If X is a subset of Y, entity
X is the hyponym of Y, and Y is the hypernym of X, then there is a hyponymy
relationship between X and Y."

At present, there are mainly the following methods about the research of
hyponymy extraction:

The pattern-based method mainly uses linguistics and natural language processing technologies to summarize the language patterns. In English, Hearst [3]

H. Li · Z. Zhang · Z. Yu (✉) · H. Wang · H. Lai
School of Information Engineering and Automation, Kunming University of Science
and Technology, Kunming, China

© Springer Nature Switzerland AG 2020   113
X. Yuan, M. Elhoseny (eds.), *Urban Intelligence and Applications*, Studies in
Distributed Intelligence, https://doi.org/10.1007/978-3-030-45099-1_9

originally put forward to extract the hyponymy relation by pattern matching in large-scale corpora. Cederberg et al. [4] extracted hyponymy by pattern matching, combined with latent semantic analysis. Snow et al. [5] adopted syntactic pattern to obtain the entity hyponymy. In Chinese, Wu et al. [6] built a general grammar pattern library for extracting the hyponymy relation. Liu et al. [7] proposed a method of acquiring hyponym entity based on the "ISA" model, analyzing the "ISA" pattern using semi-automatic dictionary and sentence pattern, and then obtaining the hyponym entity according to different rules. Their accuracy rate reached 86.3%. The pattern-based method has high accuracy, but it needs to determine manually the patterns.

Method Based on Dictionary and Knowledge Base. It makes use of existing knowledge base or dictionary to extract the hyponymy between entities. In English, Sunchanek et al. [8] used the classification information in Wikipedia to extract the semantic relations. Nakaya et al. [9] obtained the classification relationship between entities with WordNet. Lin et al. [10] exploited the semantic types in Freebase to extract. In Chinese, Fan et al. [11] made use of the Chinese Concept Dictionary, Wikipedia, Baidu Encyclopedia, Interactive Encyclopedia, and so on to extract the hyponymy, and the accuracy and recall rate were better than that of a single encyclopedia. The dictionary-based and knowledge-based method have high accuracy and recall, but it needs domain-specific dictionary as the basis.

The traditional machine learning method converts the hyponymy extraction into a classification problem. In English, Boella et al. [12] proposed a method based on the combination of dependency parsing and support vector machine to extract the hyponymy relation. In Chinese, Huang et al. [13] extracted domain terms hyponymy by making use of the conditional random fields. Wang et al. [14] adopted the conditional random fields and support vector machine to extract hyponymy. Cheng et al. [15] proposed a hybrid method for entity hyponymy acquisition in complex sentences. Conditional random fields feature template is used to identify the hyponymy entity pairs of the simple sentence, and then analysis of both the parallel relationship of entities among sentences and entity pairs in simple sentences is conducted to obtain the hyponymy entity pairs in complex sentences. However, the traditional machine learning approach requires a lot of feature engineering.

Chinese sentences are relatively complex, especially in special fields. For instance, think the following sentence: "Erhai Lake in the suburb of Dali, once known in ancient literature as Yeyuze, Kunmichuan, Xier River and Xier River, is one of the most famous scenic spots in Yunnan." Simple methods are difficult to identify accurately and traditional machine learning methods need to mark a large number of features. In recent years, many researchers have applied deep learning to relationship classification. For instance, Wang et al. [16] used the bidirectional Recurrent Neural Network for relationship classification. But the bidirectional Recurrent Neural Network has the problems of gradient disappearance and gradient explosion, which is not good for extracting long sentences. Hence, Samuel et al. [17] employed the support and evidence for implicit and explicit relationship discovery. Moreover, Zhou et al. [18] put forward attention-based bidirectional Long Short-Term Memory Networks for relation classification and experiment proved that the effect of introducing attention mechanism is better, because attention mechanism can allocate more attentions to important semantic information. However, they aim at the English datasets. There are many differences between Chinese and English sentences in experimental processing. The main reason is that Chinese syntax is complex and diverse. Deep learning has become the current mainstream method because it does not require a large number of feature engineering and greatly reduces the amount of manual work.

Therefore, this paper proposes a method of combining the bootstrapping method and attention-based bidirectional Long Short-Term Memory Networks (Bo-At-BiLSTM) for hyponymy extraction of Chinese domain complex sentences. First, the bootstrapping method is used to obtain the patterns set. Then, pattern matching is employed to acquire the candidate sentences and word embedding. Next, import into the bidirectional Long Short-Term Memory Networks and introduce attention mechanism. Finally, output the results by Softmax classifier. Bootstrapping method and pattern matching are used to improve the quality of experimental corpus. Experimental results show that the proposed method (Bo-At-BiLSTM) outperforms the baseline methods.

2 Proposed Method

As shown in Fig. 1, the proposed method of overall framework mainly includes establishment of experimental corpus, word embedding, and construction of attention-based bidirectional Long Short-Term Memory Networks (At-BiLSTM).

2.1 Extracting Candidate Sentences

2.1.1 Acquiring Patterns Set Using Bootstrapping Approach

Firstly, the text corpus crawled from the web is preprocessed by removing the duplicate information, de-noising, and so on. Next, cut the text into sentences one by one. Finally, using bootstrapping method gets patterns set. Table 1 depicts the bootstrapping approach for patterns acquisition. Table 2 shows the examples of the pattern after using bootstrapping approach.

2.1.2 Getting Candidate Sentences Using Pattern Matching

First, mark the entities in the sentences. Then, replace the entity in the sentence with E_i ($i = 1, 2, 3, ... n$). Finally, candidate sentences are obtained using pattern matching. These hyponymy candidate sentences make up the experimental corpus.

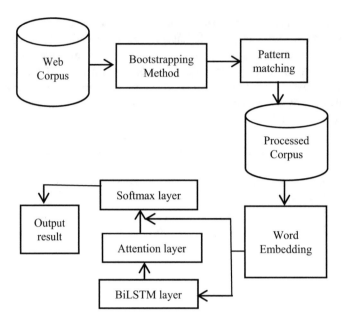

Fig. 1 Overall framework of the proposed method

Table 1 Bootstrapping approach for patterns acquisition

Input: the seeds set of hyponymy relation; Web corpus after clause
    Output: the patterns set of hyponymy relation; the number of patterns (N)
    Step:
        1. Search the seed set in corpus and get some instances S. Next, use S to generate pattern instances
        2. Summary pattern instances to get candidate patterns and add to the patterns set. The number of candidate patterns is n, $N = N + n$
        3. Match the patterns of candidate patterns in the corpus and obtain entities pairs of S as new seeds set. Add the new seeds set into seeds set
        4. Repeat the step 1, 2, 3 using new seeds set until the N no longer increases. The iteration is end and return the patterns set
---

Table 2 Examples of hyponymy patterns

Pattern	Note
(In English: E (have/mainly have/include/locate in/Specialty/the main scenic spots have/is/be situated/distribute in/:) E1, E2, ..., En) (In English: E1, E2,..., En (compose of / and so on) E)	E is the domain entity E1, E2, ..., En are entity instances

2.2 Word Embedding

Word embedding is used to translate words into dense vectors that can be understood by computers and convert sentences into a multidimensional matrix. Given a sentence composing of n words $S = \{w_1, w_2, w_3, \ldots w_n\}$, each word w_i is mapped into a low-dimensional vector space by using word2vec [19]. Formula (1) gives the word vector processing of text information:

$$r^w = w^{\text{word}} \times v^w \tag{1}$$

r^w is the word vector representation of the word w. w^{word} stands for the word vector matrix. v^w is the one-hot representation of the word.

Given the features set $K = (K_1, K_2, K_3, \ldots K_n)$ of sentence S, the i-th feature of the t-th word is expressed as $w_t^{k_i}$. Formula (2) gives the word vector processing of text feature:

$$r^{k_i} = w^{k_i} \times v^w \tag{2}$$

r^{k_i} is the word vector representation of the i-th feature of the text. w^{k_i} is the feature distribution vector of the i-th feature of the text, $w^{k_i} = \left(w^{k_1}, w^{k_2}, w^{k_3}, \ldots, w^{k_n} \right)$.

The vectorization of every word is the connection of each vector. Formula (3) gives the vectorization of the i-th word:

$$x_i = \left[r_i^w, r_i^{k1}, r_i^{k2}, r_i^{k3}, \ldots, r_i^{kn} \right] \tag{3}$$

Finally, output of the sentence S in the word embedding layer is Formula (4).

$$\text{emb} = \{x_1, x_2, x_3 \ldots x_n\} \tag{4}$$

2.3   Bidirectional Long Short-Term Memory Network

Bidirectional Long Short-Term Memory Network (BiLSTM) combines Bidirectional Recurrent Neural Network (BRNN) and LSTM units. The LSTM unit consists of three "gates" and a memory cell when the information comes to the LSTM unit, it will pass the "forget gate," "input gate," and "output gate" to realize the discarding and updating of information. The "forget gate" determines which information will be discarded. The "input gate" determines which information will be joined. The "output gate" outputs the information after merging and updating. This design gives the LSTM cells the ability to read, reset, save, and update long-range information. Thus, LSTM can be used to deal with domain complex sentences.

BiLSTM considers the context of the text and make full use of the whole information of the text. In BiLSTM, forward direction ($\overrightarrow{h}_t$) is responsible for obtaining the above information and backward direction ($\overleftarrow{h}_t$) is in charge of the below information. Finally, forward direction and backward direction are linked to the same output.

$$\overrightarrow{h_t} = \overrightarrow{LSTM}_{(h_{t-1}, w_t, c_{t-1})}, t \in [1, T] \tag{5}$$

$$\overleftarrow{h_t} = \overleftarrow{LSTM}_{(h_{t+1}, w_t, c_{t+1})}, t \in [T, 1] \tag{6}$$

$$H_t = \left[\overrightarrow{h_t}, \overleftarrow{h_t} \right] \tag{7}$$

2.4 Attention Layer

Attention layer is to introduce attention mechanism. The attention mechanism simulates the attentional characteristics of the human brain and distributes more attentions for important information in the text. Attention mechanism has been widely used in natural language processing. Chorowski et al. [20] used the attention mechanism for speech recognition. Lu et al. [21] made use of the attention mechanism to classify long texts. Bahdanau et al. [20] adopted the attention mechanism for machine translation. There are hidden units u and attention vector a in the attention mechanism.

$$u_t = \tanh(w_w H_t + b_w) \tag{8}$$

$$a_t = \text{softmax}\left(u_t^T, u_w\right) \tag{9}$$

$$v = \sum_t a_t H_t \tag{10}$$

where u_t is the hidden unit of H_t, and b_w represents attentional bias vector. The u_w stands for the context vector; a_t is the attention vector; v represents the output vector after introducing the attention mechanism.

2.5 Softmax Layer

The Softmax layer mainly includes the following steps:

1. Pooling: the output features of Attention layer are pooled by max-pooling method to obtain the whole features of text with fixed dimensions.
2. Feature fusion: feature fusion to get new feature.
3. Output results: import new feature into the Softmax classifier for the classification and output results.

Table 3 entity hyponymy main class and subclass for tourism field

Entity hyponymy main class	Entity hyponymy subclass
Geographical location relation	Locate in, be situated, distribute in
Structural affiliation relation	Belong to, Specialty, Scenic spot, have, is

3 Experiments

3.1 Experimental Data

The experimental corpus of this paper contains more than 20,000 pieces of travel domain texts crawled from encyclopedic entries and travel websites. Patterns matching yields 10,000 candidate sentences, 8000 as training data, and 2000 as test data. This paper combines the characteristics of hyponymy and Lei et al. [22] definition of the relationship of Chinese tourism domain giving the two entities entity hyponymy main class (Table 3).

3.2 Parameter Settings

The experiment selected the Tensorflow framework developed by Google as the training tool of model. Relu function is activation function. The nodes of hidden layer are 300. In order to prevent over-fitting, introduce dropout and set the dropout rate 0.5. The optimization method uses Adadelta with batch size of 50 and training rounds of 500. Word vector training makes use of the Google Open Source Toolkit Word2vec. The window size is 5. The vector dimension is 200 dimensions.

3.3 Evaluating Indicator

This article uses the most common precision (*P*), recall (*R*), and F-Measure value (*F*) as the evaluation criteria. The calculation formula is as follows:

$$P = \frac{A}{B} \times 100\% \tag{11}$$

$$R = \frac{A}{C} \times 100\% \tag{12}$$

$$F = \frac{2 \times P \times R}{P + R} \times 100\% \tag{13}$$

where *A* represents the correct number of the extracted domain entities hyponymy; *B* represents the total number of domain entities hyponymy, *C* represents the total number of domain entities hyponymy in the test set.

3.4   Experimental Design

In order to verify the feasibility of the proposed method, the following three experiments are set up in this paper.

Experiment 1: First, the Web corpus is preprocessed by de-duplication, de-noising, and clause processing. Then bootstrapping method is used to get patterns set. Finally, candidate sentences are extracted by pattern matching.

Experiment 2: The processed corpus after pattern matching is the corpus of Experiment 2. Using the proposed method (Bo-At-BiLSTM), CRF, LSTM, and BiLSTM extract the hyponymy relation of tourism field.

Experiment 3: The Web corpus after being preprocessed by de-duplication, de-noising, and clause processing is used as the experimental data of Experiment 3. Using the At-BiLSTM model constructed in this paper extracts the hyponymy relation of tourism field. The rest of the parameter settings are the same as in Experiment 2.

Experiment 4: Experiment 4 was conducted using the SemEval-2010 Task8 [23] public corpus. The sample relationship is shown in Table 4. The relationship extraction is performed using the At-BiLSTM model of this paper. The parameter settings are the same as in Experiment 3.

3.4.1   Experimental Results and Analysis

Results of experiment 1: Table 5 shows the experimental results of pattern matching.

Results of experiment 2: Table 6 shows the experimental results of the proposed approach. Table 7 shows the comparison of the results from the different approaches.

Table 4 Sample relationship table

Relationship type	Number of samples
Cause-effect	1331
Component-whole	1253
Entity-destination	1137
Entity-origin	974
Product-producer	948
Message-topic	895
Content-container	732
Instrument-agency	660
Other	1864

Table 5 Experimental results of pattern matching

Algorithm	P	R	F
Pattern matching	64.17	81.30	71.52

Table 6 Experimental results of the proposed method

Types of hyponymy relations	P	R	F
Geographical location relation	81.05	81.91	81.48
Structural affiliation relation	80.04	80.34	80.17
Other	75.20	76.84	75.50
All	80.19	83.04	82.06

Table 7 Experimental results of various methods

Model	P	R	F
CRF	70.43	69.83	70.04
LSTM	76.24	75.03	75.18
BiLSTM	78.13	80.62	78.46
Bo-At-BiLSTM	80.19	83.04	82.06

Table 8 Experimental results of experiment 3

Types of hyponymy relations	P	R	F
Geographical location relation	70.12	71.70	72.36
Structural affiliation relation	70.10	70.24	70.09
Other	76.31	77.75	76.50
All	69.19	72.04	71.06

It can be seen from Table 6 that the extraction effect of simple geographical location relationship is better than structural affiliation relation. Simple semantic relationship is easier to learn by model.

Table 7 shows that:

1. The deep learning method is better than the traditional machine learning method.
2. BiLSTM is better than LSTM because BiLSTM can make full use of context information.
3. The effect of the proposed method is better than BiLSTM because it introduces the attention mechanism. The attention mechanism imitates the attentional characteristics of the human brain. It can allocate more attentions for important information

Table 8 shows the experimental results of experiment 3.

By comparing Tables 6 and 8, it can be seen that the extraction effect of Table 6 is much better than Table 8. The main reason is that the quality of corpus is different. The corpus of experiment 3 did not use pattern matching for filtering. The corpus of Experiment 2 first acquires patterns set through bootstrapping methods, and then gets the candidate sentences by pattern matching. These candidate sentences are the sentences with hyponymic patterns, which greatly improves the quality of the corpus.

Table 9 shows the experimental results of experiment 4. By comparing the results of Experiment 3 with Table 8 and the results of Experiment 4, Table 9, it is apparent

Table 9 Experimental
results of experiment 4

Relationship type	P	R	F
Cause-effect	79.16	88.61	83.79
Component-whole	89.31	89.63	89.59
Entity-destination	89.96	92.36	92.14
Entity-origin	83.03	85.33	84.16
Product-producer	91.75	92.53	92.15
Member-collection	93.45	96.46	86.47
Message-topic	86.15	87.01	86.47
Content-container	89.07	89.47	89.21
Instrument-agency	82.58	83.31	83.01
Other	79.30	81.94	80.60
All	85.90	88.44	87.16

that the overall value of Table 9 is higher than Table 8. The main reason is that the corpus of Experiment 3 and Experiment 4 are different. Experiment 3 is a Chinese travel corpus. Experiment 4 is the SemEval-2010 Task8 English Public Corpus. Chinese needs word segmentation and complex and diverse syntactic forms are more difficult to deal with than English. The good results of Experiment 4 also indicate that the At-BiLSTM model is suitable for processing entity relationship extraction tasks.

4 Conclusions

This paper proposes a method that combines Bootstrapping method and Attention-Based Bidirectional Long Short-Term Memory Networks (Bo-At-BiLSTM) to extract the field entity hyponymy of complex sentences. First, the bootstrapping method is used to obtain the patterns set. Then, pattern matching is employed to acquire the candidate sentences and word embedding. Next, import into the bidirectional Long Short-Term Memory Networks and introduce attention mechanism. Finally, output the results by Softmax classifier. It can be seen from the experimental results that this method performs well in the entity hyponymy extraction of complex sentence. Through comparative analysis, it is known that using the bootstrapping method can actually improve the quality of corpus and then make the extraction result more accurate. It proves that the proposed method is feasible.

Acknowledgments This paper is supported by National key research and development plan project (Grant Nos. 2018YFC0830105, 2018YFC0830100), National Natural Science Foundation of China (Grant Nos. 61732005, 61672271, 61761026, and 61762056), Yunnan high-tech industry development project (Grant No. 201606), Natural Science Foundation of Yunnan Province (Grant No. 2018FB104), and National Natural Science Foundation of China (Grant Nos. 61562052, 61462054, and 61866019).

References

1. X. Yuan, D. Ang, A novel figure panel classification and extraction method for document image understanding. Int. J. Data Min. Bioinform. **9**(1), 22–36 (2014)
2. G.A. Miller, WordNet: a lexical database for English. Commun. ACM **38**(11), 39–41 (1995)
3. M.A. Hearst, Automatic acquisition of hyponyms from large text corpora, in *Proceedings of the 14th Conference on Computational Linguistics-Volume 2*, (Association for Computational Linguistics, Stroudsburg, 1992), pp. 539–545
4. S. Cederberg, W. Dominic, Using Isa and noun coordination information to improve the precision and recall of automatic hyponymy extraction. in *Proceedings of the Seventh Conference on Natural Language Learning at Human Language Technology & North American Chapter of the Association for Computational Linguistics*, pp. 111–118, 2003
5. R. Snow, D. Jurafsky, A.Y. Ng, Learning syntactic patterns for automatic hypernym discovery. Adv. Neural Inf. Proces. Syst. **17**, 1297–1304 (2004)
6. J. Wu, C.C. Robert, et al., Acquisition and validation of some relational knowledge in web pages. J. East China Univ. Technol. **8**, 11 (2006)
7. L. Liu, C.G. Cao, H.T. Wgng, et al., A method of hyponym acquisition based on 'isa' pattern. Comput. Sci. **33**(9), 146–151 (2006)
8. F.M. Suchanek, G. Kasneci, G. Weikum, Yago: a large ontology from wikipedia and WordNet. J. Web Semant. **6**(3), 203–217 (2008)
9. N. Nakaya, M. Kurematsu, T. Yamaguchi, A domain ontology development environment using a MRD and text corpus, in *Proc of the Joint Conf on Knowledge Based Software Engineering*, (IOS Press, Amsterdam, 2002), pp. 242–253
10. T. Lin, O. Etzioni, No noun phrase left behind: detecting and typing unlink able entities. in *Proceedings of the 2012 Joint Conference on Empirical Methods in Natural Language Processing and Computational Natural Language Learning*, pp. 893–903, 2012
11. Q.H. Fan, H.Y. Zan, Y.M. Chai, et al., Hyponym discovery of multiple resource fusion. Comput. Eng. Des. **34**(12), 4310–4315 (2013)
12. G. Boella, L.D. Caro, Extraction definitions and hypernym relations relying on syntactic dependencies and support vector machines. in *Proceedings of the 51st Annual Meeting of the Association for Computational Linguistics*, pp. 532–537, 2013
13. Y. Huang, Q. Wang, Y. Liu, An acquisition method of domain-specific terminological hyponymy based on CRF. J. Cent. South Univ. **44**(2), 355–359 (2013)
14. C. Wang, Z. Yang, A method of acquiring domain terms based on sentence structure features. J. Chongqing Univ. Posts Telecommun. **26**(3), 385–389 (2014)
15. Y. Cheng, J. Guo, Y. Xian, A hybrid method for entity hyponymy acquisition in Chinese complex sentences. Autom. Control. Comput. Sci. **50**(5), 369–377 (2016)
16. D. Zhang, D. Wang, Relation classification via recurrent neural network. arXiv preprint arXiv:1508. 01006, 2015
17. J. Samuel, X. Yuan, X. Yuan, B. Walton, Mining online full-text literature for novel protein interaction discovery. in *2010 IEEE International Conference on Bioinformatics and Biomedicine*, pp. 277–282, 2010
18. P. Zhou, W. Shi, J. Tian, et al., Attention-based bidirectional long short-term memory networks for relation classification. in *Proceedings of the 54th Annual Meeting of the Association for Computational Linguistics, Berlin, Germany, August 7–12*, pp. 207–212, 2016
19. T. Mikolov, I. Sutskever, C. Kai, et al., Distributed representations of words and phrases and their compositionality. 2014. https://papers.nips.cc/paper/5021-distributed-representations-of-words-and-phrases-and-their-compositionality.pdf
20. J.K. Chorowski, D. Bahdanau, D. Serdyuk, K. Cho, Y. Bengio, Attention-based models for speech recognition. in *Advances in Neural Information Processing Systems*, pp. 577–585, 2015
21. L. Lu, Y. Wu, Y. Wang, et al., Long text categorization combined with attention mechanism. Comput. Appl. **38**(5), 1272–1277 (2018)

22. C.Y. Lei, J.Y. Guo, Z.T. Yu, et al., The field of automatic entity relation extraction based on binary classifier and reasoning, in *Third International Symposium on Information Processing* (IEEE, Piscataway, 2010)
23. M.T. Luong, H. Pham, C.D. Manning, Effective approaches to attention-based neural machine translation. Comput. Therm. Sci. **2015**, 11 (2015)

Emotion Recognition Based on EEG Signals Using LIBSVM as the Classifier

Tian Chen, Sihang Ju, Fuji Ren, Mingyan Fan, and Xin An

Abstract In order to improve the emotion recognition rate, this paper proposes an electroencephalograph (EEG) emotion recognition model using a library for support vector machine (LIBSVM) as the classifier. In this paper, we collected EEG signals from 10 volunteers along with the participants' self-assessment of their affective state after each stimulus, in terms of Valence and arousal. After these signals are filtered, we calculate the features of Lempel-Ziv complexity, wavelet detail coefficient, and the co-integration relationship degree first. At the same time, EMD is carried out to calculate the average approximate entropy of the first four Intrinsic Mode Functions (IMFs). At last, all the features extracted will input into the LIBSVM for training and testing, and complete emotion recognition. In this paper, two classifications are carried out on the two dimensions of Valence and Arousal, respectively. The experimental results show that the average emotional recognition rate is 83.64% and 75.11%, respectively, which proves that the proposed scheme has a certain feasibility.

Keywords Text Mining · Entity Linking · Topic Model · Latent Dirichlet Allocation · Semantic

T. Chen (✉) · S. Ju · M. Fan · X. An
School of Computer Science and Information Engineering, Hefei University of Technology, Hefei, Anhui, China

Anhui Province Key Laboratory of Affective Computing and Advanced Intelligent Machine, Hefei University of Technology, Hefei, Anhui, China
e-mail: ct@hfut.edu.cn

F. Ren
School of Computer Science and Information Engineering, Hefei University of Technology, Hefei, Anhui, China

Anhui Province Key Laboratory of Affective Computing and Advanced Intelligent Machine, Hefei University of Technology, Hefei, Anhui, China

Faculty of Engineering, The University of Tokushima, Tokushima, Japan

1 Introduction

Emotion is a comprehensive physiological and psychological state of people's various feelings, thoughts, and behavioral states. It is the attitude experience of people on whether objective things meet their needs a psychological and physiological response of humanity to external stimuli. With the rapid rise of natural human–computer interaction technology and artificial intelligence, emotions have begun to be studied, such as facial expression-based emotion recognition [1], intelligent control based on the emotional state [2, 3], emotion-based image retrieval [4], etc. At the same time, emotion recognition has certain application value in the fields of emotion regulation and disease detection [5].

Among the research and application of emotion, high-efficiency emotion state recognition is the foundation and key. In human–computer interaction, in order to make robots possess the ability to perceive and understand human emotions, we must improve the accuracy of emotion recognition. The traditional method of emotion recognition is often judged by external expressions and behavioral actions, but sometimes, people's external behavior may be deliberately disguised, which has certain interference to the identification of the real inner emotion. In recent years, with in-depth research in Neuroscience and the continuous development of science and technology, electrophysiological signals have begun to be used for emotion recognition and classification. Such as electroencephalograph (EEG), electrocardiograph (ECG), electromyography (EMG), and eye movement signal (EOG). Among them, emotional recognition based on EEG signals, in the field of emotion recognition, the experimental results are ideal [6].

At present, for emotion classification, there are often two emotional models. One is the discrete emotional model that contains happiness, anger, fear, hate, sadness, and surprise [7]. The other is the continuous sentiment model, usually using a two-dimensional emotional model of Arousal and Valence [8] and a VAD three-dimensional emotional model [9]. In order to identify the emotional state, different strategies can be used to quantify the EEG signal data, and then establish the mathematical model of the corresponding emotion for emotion analysis.

In the process of emotion recognition, emotion classification based on the EEG has been paying attention. Wen et al. [10] propose an end-to-end model which is based on Convolutional Neural Networks (CNNs). In order to represent the EEG signals better, the original channels of EEG are firstly rearranged by Pearson Correlation Coefficient and the rearranged EEGs are fed into CNN. The experimental results achieve 77.98% accuracy on the Valence recognition and 72.98% on the Arousal recognition. Zheng et al. [11] use deep belief networks (DBNs) to construct EEG-based emotion recognition models for three emotions: positive, neutral, and negative. They selected four different profiles to include 4, 6, 9, and 12 channels, and the recognition accuracies of these four profiles are relatively stable with the best accuracy of 86.65%. The result is even better than using the original 62 channels and proves the importance of channel selection. Putra et al. [12] use Wavelet Decomposition and k-Nearest Neighbor (kNN) to improve accuracy. When k is 21, the result of Valance and Arousal classification accuracy is 57.5% and 64.0%. And an approach on recognizing electroencephalography (EEG) emotion using empirical wavelet transform (EWT) and autoregressive (AR) model is given in the literature [13].

In order to further improve the emotion recognition rate, this paper combines empirical mode decomposition (EMD) and average approximate entropy and performs EMD on the wavelet-filtered EEG signal to calculate the average approximate entropy of the first four Intrinsic Mode Functions (IMFs). It is combined with the Lempel-Ziv complexity, wavelet detail coefficient, and the co-integration relationship degree that is proposed in this paper according to the co-integration test that are calculated for the EEG after wavelet filtering as the features. Using a library for support vector machine (LIBSVM) as the classifier for training and testing.

The structure of the remaining part of this paper is as follows: the second part describes the overall process of establishing the EEG emotion recognition model using LIBSVM as the classifier; the third part describes the experimental process of this paper; the fourth part analyzes the experimental results. The fifth part summarizes the work of this paper.

2 EEG Emotion Recognition Model Using LIBSVM the Classifier

2.1 The Overall Framework of the Proposed Emotion Recognition Model

This paper proposes an EEG emotion recognition model using LIBSVM as the classifier for emotion recognition. The overall framework is shown in Fig. 1. It is mainly divided into the following four steps:

Step1: The EEG signal obtained by the video excitation is subjected to filtering processing. In this paper, the Butterworth filter is used for filtering to obtain EEG signals with a frequency of 1–43 Hz.

Step2: The Lempel-Ziv complexity, wavelet detail coefficient, and the co-integration relationship degree are calculated for the filtered EEG signal. The process is the feature extraction before EMD.

Step3: For the filtered EEG signals, they are decomposed using EMD, and the average approximate entropy is calculated for the first four IMFs. The process is the feature extraction after EMD.

Step4: The feature values calculated in steps 2 and 3 are combined and sent to the LIBSVM classifier for training and testing, and finally emotion recognition is performed.

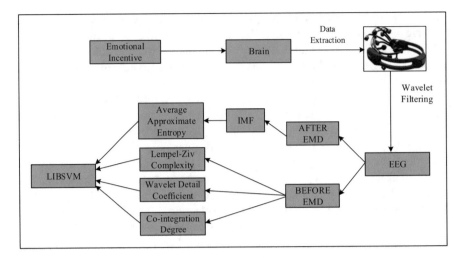

Fig. 1 The overall flow of the EEG emotion recognition model

In this paper, different subjects are selected to participate in the process of the experiment. Different channel combinations are also studied to participate in the process of emotional recognition.

2.2   Feature Extraction Before EMD Decomposition

Lempel-Ziv Complexity
The Lempel-Ziv complexity can explore the dynamic complexity of brain activity [14] and reflect the time domain information of the signal. For each sample, the Lempel-Ziv complexity of each channel is calculated.

Wavelet Detail Coefficient
The wavelet detail coefficient reflects the frequency domain information of the signal. Under each channel, the data in each second is separately decomposed using db5 wavelet, and the detail coefficients of 1, 2, and 3 are reconstructed. For each reconstructed wavelet coefficient, calculate the mean value of the wavelet detail coefficient obtained per second, and the mean value of the detail coefficient in each second is added, divided by the acquisition time, which is the final wavelet detail coefficient.

Co-integration Relationship Degree
Co-integration test determines whether there is a certain equilibrium relationship between non-stationary sequences, and reflects whether there is a relationship between signals under different channels. For each of the two channels under all channels, use the EG test [15] to check whether there is a co-integration relationship

between the data of the two channels per second, and count the total number of co-integration relationships in each time period. Dividing the total number of co-integration relationships of the two channels by the acquisition time, the result obtained is referred to herein as the co-integration relationship quantity between the two channels.

On this basis, for each channel, we select an optimal channel from other channels, and the co-integration relationship quantity between the channel and the selected optimal channel is called the co-integration relationship degree of the channel. The selected optimal channel is called the co-integration optimal relation channel. ReliefF algorithm [16] is used to select the optimal channel, and the specific selection process is as follows:

Step1: For each category, randomly select 1/3 samples, and the total number is recorded as m. And take the co-integration relationship quantity between the two channels as a feature, use w_{ij} to represent its weight (i, j represents the channel position, and i is not equal to j, $i = 1 \sim n$, $j = 1 \sim n$, n is the total number of channels), all the weights are initialized to 0.

Step 2: For calculating the co-integration relationship quantity of channel i, take j from 1 to n, and j does not take i. For each channel j selected, for each sample selected, the modified formula of weight can be obtained as shown in (1)–(3).

$$z1 = diff(d_{ij}, h_{ij})/m \qquad (1)$$

$$z2 = P(C)/(1 - P(class(R))) * diff(d_{ij}, f_{ij})/m \qquad (2)$$

$$w_{ij} = w_{ij} - z1 + z2 \qquad (3)$$

Among them, d_{ij} is the co-integration relationship quantity of the selected sample between channel i and channel j, h_{ij} is the co-integration relation quantity nearest to d_{ij} between channel i and channel j of the same sample; f_{ij} is the co-integration relation quantity nearest to d_{ij} between channel i and channel j of heterogeneous sample. $P(C)$ is the ratio of the category, and $P(class(R))$ is the ratio of the selected sample to a certain category, both of which have a value of 0.5. Among them, the calculation process of the function diff is as shown in Eq. (4), max and min represent the maximum value corresponding to this feature and the minimum value corresponding to this feature, respectively.

$$diff(y1, y2) = abs(y1 - y2)/(max - min) \qquad (4)$$

Step3: After all the samples have been processed by formula 2, for channel i, the weight between all other channels can be obtained. If channel j with the largest weight is selected, then channel j is the co-integration optimal relationship channel, and the co-integration relationship quantity between two channels is the co-integration relationship degree of channel i.

2.3   Feature Extraction After EMD Decomposition

EMD is a time-frequency analysis method for nonlinear and unsteady signals [17]. It can decompose complex signals into a finite number of IMFs, and the decomposed IMF components contain local characteristic signals of different time scales of the original signal. The IMF must satisfy the following two conditions:

1. During the entire data period, the extreme points the number of zeroes and the number of zero crossings must differ no more than one;
2. At any moment, the average of the local upper and lower envelopes is 0.

On this basis, for the original signal $x(t)$, the decomposition process of EMD can be divided into the following steps:

Step1:   Find out all the maximum and minimum points of the original data sequence $x(t)$, and then obtaining the upper envelope and lower envelope of the signal using the cubic spline interpolation. The mean of the upper and lower envelopes is $m(t)$;

Step2:   Use $x(t) - m(t)$ to get $n(t)$, and check whether $n(t)$ satisfies the condition of the IMF. If not, continue to decompose $n(t)$ and go to step (1); if satisfied, go to step (3);

Step3:   At this time, $n(t)$ is the IMF, and the IMF is subtracted from $x(t)$ to obtain $r(t)$. If $r(t)$ can be decomposed, go to step (1) to continue the decomposition; if it cannot be decomposed, the decomposition ends.

The EMD decomposition of the signal $x(t)$ is finally shown in (5).

$$x(t) = \sum_{i=1}^{n} \text{IMF}_i \tag{5}$$

For the filtered EEG signals, they are decomposed using EMD and the first four IMFs are selected for approximate entropy calculation. Approximate entropy is a method to measure the complexity and regularity of time series [18]. It is to distinguish the complexity of the time process from a statistical point of view, and relies on a less amount of data to calculate and is robust to noise [19].

In the calculation of the approximate entropy, two parameters z and r need to be input. z is called the mode dimension, which is the length of the comparison sequence, that is, the window length; r is the tolerance of the similarity between the sub-segments, and the size of r is generally set to 0.2–0.3. For each IMF, calculating the approximate entropy of the signal per second. Then accumulate the approximate entropy per second of each IMF, divided by the acquisition time, and the average approximate entropy is obtained.

2.4 LIBSVM

After the above feature extraction process, the Lempel-Ziv complexity, wavelet detail coefficient, the co-integration relationship degree and the average approximate entropy of the first four IMFs will be obtained. These features will be combined into a feature vector that will eventually be used as the feature input to the classifier.

This article uses LIBSVM as the classifier for training and testing. For each EEG channel signals, corresponding feature values are extracted to participate in the process of emotion recognition, and different channels are independent of each other. For the channels involved in emotion recognition, voting is used to determine the final emotional recognition result. In the training, this paper considers all the channel combinations, and selects the channel combination with the best average emotion recognition rate to test and perform emotion recognition.

3 Experimental Process

3.1 Experimental Equipment

Experimental equipment is the Emotiv Epoc electroencephalograph, which has 16 electrodes for the experimenter to wear. The electrode diagram is shown in Fig. 2. Among them, CMS and DRL are two reference electrodes, and the remaining 14 electrodes are used to collect EEG signals.

3.2 Experimental Objects and Experimental Materials

The correct choice of the experimental object will determine the reliability of the experimental data. The subjects are undergraduate and graduate students, without any history of cranial nerve injury and mental illness, and have better sleep quality.

Fig. 2 Emotiv Proc electrode position map

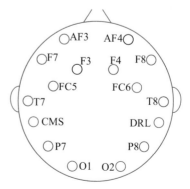

In the experiment of stimulating emotional state, researchers usually use pictures [20], music [21], and video [22]. In order to better stimulate the emotional state, this article chooses five kinds of video material: glad, relaxed, angry, sad, and disgusting. First, we select some material under each emotional state from the video libraries. Then let some volunteers choose one of the strong, medium, and weak strengths after watching these videos. Finally, video material with strong or medium degree is selected for the experiment. In the end, for each kind of emotion, we select 8 videos to totaling 40 videos. There are 8 men and 2 women participated in the experiment to watch these videos to collect EEG signals.

3.3   Emotional Incentive

When subjects conduct the experiment, the specific experimental process is shown in Fig. 3. Each subject needs to watch 40 videos. Subjects watch the first video need to focus on 5s, and then there will be a text reminder of 5s. After watching each video, the emotional state of the video is evaluated. After the evaluation, there will be a rest time of about 1 min to alleviate and eliminate the influence of the previous video. After that, the text reminder of 5s will continue, and the operation will be repeated until 40 videos are all viewed.

3.4   Emotion Assessment

In the process of the experiment shown in Fig. 3, the subject performs a corresponding emotion assessment after viewing each video. Subjects scored on the Valance and Arousal dimensions, respectively, with a range of 1–9 points.

4   Experimental Results

After the above experimental process, the EEG data under the Valance and Arousal dimensions are obtained. This paper classifies emotions by scores, which is higher

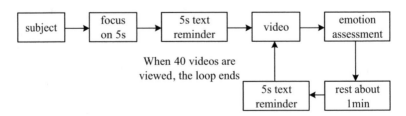

Fig. 3 The overall flow of the EEG emotion recognition model

than 5 into one category, less than 5 into one category, and each dimension is classified into two categories. At this time, the cross-validation strategy is adopted to group the data, and half of the data is used as a training set to establish the EEG emotion recognition model using LIBSVM as the classifier; the other half of the data is used as a verification set, and finally the recognition rate of emotion classification is obtained.

In the process of training, this paper considers the combination of all channels, and selects the channel combination with the best average emotion recognition rate for testing. The combination of channels obtained for testing in this way is the optimal channel combination. In the process of training, the optimal channel combination of Valance and Arousal are shown in Fig. 4a and b, respectively. When using the channel combination of Fig. 4a and b for testing, the highest average emotion recognition rate is 83.64% and 75.11% for Valance and Arousal, respectively. At this time, the kernel function of LIBSVM is a Gaussian kernel function with a penalty factor of 1. It can be seen from the experimental results that the average emotion recognition result for Valance is better than Arousal.

Based on selecting the optimal channel combination, another four channel combinations are selected and compared with it, as shown in Fig. 5. Select the four channel combinations in Fig. 5 to participate in the process of emotion recognition and the final experimental results are shown in Fig. 6. It can be seen from Fig. 6 that the emotion recognition rate of the optimal channel combination is higher than the other four strategies.

In order to verify the feasibility of Gaussian kernel function, this paper chooses linear kernel function, polynomial kernel function, and perceptron kernel function, and selects the channel combination with the optional average emotional recognition rate under training under the kernel function to perform emotion recognition. The results are shown in Fig. 7. It can be seen from Fig. 7 that the Gaussian kernel function has the optimal average emotion recognition rate.

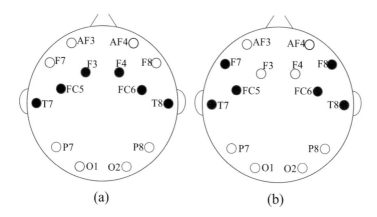

Fig. 4 Channel combination Valance and Arousal with optimal average emotion recognition rate

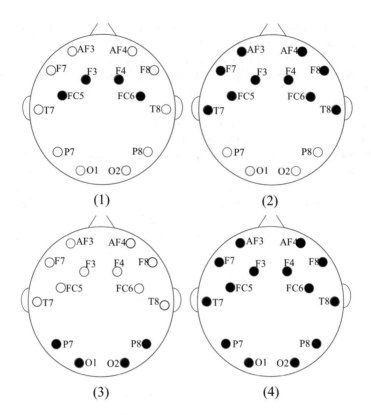

Fig. 5 Comparison channel combinations

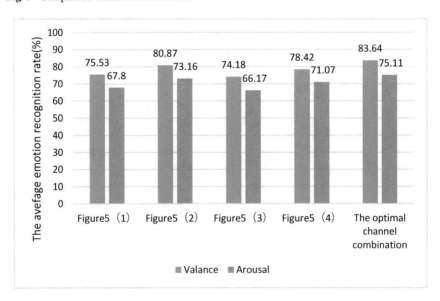

Fig. 6 Comparison results

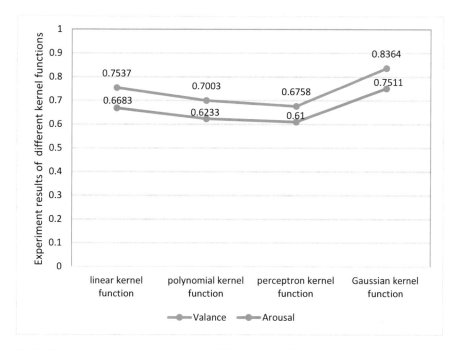

Fig. 7 Comparison of experimental results of different kernel functions

Table 1 Comparison of average sentiment recognition rates with other programs

Classifier	CNN [10]	KNN [12]	EWT [13]	LDA [23]	Proposed method
Valance	77.98%	57.5%	64.3%	79.06%	83.64%
Arousal	72.98%	64.0%	67.3%	77.19%	75.11%

Table 1 shows the comparison of the scheme of this paper with some existing schemes in emotion recognition. Wen et al. [10] achieve an average recognition rate for Valance and Arousal at 77.98% and 72.98%, respectively. And the recognition rate for Valance and Arousal in [12] is 57.5% and 64.0%. Huang et al. [13] combine empirical wavelet transform (EWT) with autoregressive (AR) model, and the approach achieves 64.3% for Valance and 67.3% for Arousal. Xie et al. [23] propose a stacking emotion classification, in which different classification models such as XGBoost, LightGBM, and Random Forest are integrated to learn the features. The result shows the average recognition accuracies of 77.19% for Arousal and 79.06% for Valance. The average accuracy rate of this paper's method for Valance and Arousal is at 83.64% and 75.11% and is higher than the method in [10, 12] and [13]. Although for Arousal, the average accuracy rate in literature [23] is higher than the paper's method, the average accuracy rate for Valance and Arousal of this paper's method is 79.375% and in [23] is 78.125% by calculating. So, the classification performance of this paper's method is better than the method in [23].

5 Conclusion

This paper proposes an EEG emotion recognition model using LIBSVM as the classifier, which implements two classifications in two dimensions: Valance and Arousal, respectively. In the process of emotion recognition, not all channels are beneficial to emotion recognition. Different channel combinations participate in the emotion recognition process and their recognition results are different. In the choice of kernel function of LIBSVM, this paper studies the effects of linear kernel function, polynomial kernel function, perceptron kernel function, and Gaussian kernel function on emotion recognition rate. Four kernel functions were used in testing and the training was conducted with channel combination using the best average emotion recognition rate. When the Gaussian kernel function is selected, the average emotion recognition rate of the proposed scheme for Valance can reach 83.64% and for Arousal can reach 75.11%, that is, the average emotion recognition rate reaches the highest.

In the future work, how to further improve the average emotion recognition rate of the two classifications in the two dimensions of Valance and Arousal, respectively, will be studied. At the same time, the multi-classification of emotion dimension is also the focus of the next work. On this basis, how to integrate other physiological signals to improve the rate of emotional recognition is also a direction of future research.

Acknowledgments This work supported by The Key Program of the National Natural Science Foundation of China (Grant No. 61432004); The National Natural Science Foundation of China (Grant No. 61474035, 61502140); The fund of Affective Computing and Advanced Intelligent Machine Anhui Province Key Laboratory (Grant No. ACAIM180101); NSFC-Shenzhen Joint Foundation (Key Project) (Grant No. U1613217).

References

1. Y. Lee, W. Han, Y. Kim, Emotional recognition from facial expression analysis using Bezier curve fitting, in *2013 16th International Conference on Network-Based Information Systems* (2013)
2. M. Ogino, Y. Mitsukura, An EEG-based robot arm control to express human emotions, in *2018 IEEE 15th International Workshop on Advanced Motion Control (AMC)* (2018)
3. B. Fang, Q. Zhang, H. Wang, X. Yuan, Personality driven task allocation for emotional robot team. Int. J. Mach. Learn. Cybernet. **9**(12), 1955–1962 (2018)
4. Y. Kim, Y. Shin, S. Kim, E. Kim, H. Shin, Emotion-based image retrieval, in *International Conference on Consumer Electronics (IEEE)* (2009)
5. T. Chen, S. Ju, X. Yuan, M. Elhoseny, F. Ren, M. Fan, Z. Chen, Emotion recognition using empirical mode decomposition and approximation entropy. Comput. Electric. Eng. **72**, 383–392 (2018)
6. V. Bajaj, R.B. Pachori, Detection of human emotions using features based on the multiwavelet transform of EEG signals, in *Brain-Computer Interfaces* (Springer International Publishing, Cham, 2015), pp. 215–240

7. P. Ekman, Contacts across cultures in the face and emotion. J. Pers. Soc. Psychol. **17**(2), 124–129 (1971)
8. J.A. Russell, A circumplex model of affect. J. Personal. Soc. Psychol. **39**(6), 1161 (1980)
9. M.M. Branley, P.J. Lang, Measuring emotion: the self-assessment manikin and the semantic differential. J. Behav. Therapy Exp. Psychiatr. **25**(1), 49–59 (1994)
10. Z. Wen, R. Xu, J. Du, A novel convolutional neural networks for emotion recognition based on EEG signal, in *2017 International Conference on Security, Pattern Analysis, and Cybernetics (SPAC)* (2017)
11. W.L. Zheng , B.L. Lu, Investigating critical frequency bands and channels for EEG-based emotion recognition with deep neural networks. IEEE Trans. Autonom. Mental Develop. **7**(3), 1–1 (2015)
12. A.E. Putra, C. Atmaji, F. Ghaleb, EEG-based emotion classification using wavelet decomposition and K-nearest neighbor, in *2018 4th International Conference on Science and Technology (ICST)* (2018)
13. D. Huang, S. Zhang, Y. Zhang, EEG-based emotion recognition using empirical wavelet transform, in *2017 4th International Conference on Systems and Informatics (ICSAI)* (2017)
14. Y. Wang, N. Wang, Y. Li, H. Li, J. Wang, Abnormal sensory gating in schizophrenia: the evidence from Lempel-Ziv complexity analysis in EEG, in *International Conference on Biomedical Engineering & Informatics* (IEEE, New York, 2011)
15. C.W. Granger, Investigating casual relations by econometric model and cross-spectral model. Econometrica **37**, 424–438 (1969)
16. M. Wu, Y. Wang, A feature selection algorithm of music genre classification based on ReliefF and SFS, in *IEEE/ACIS International Conference on Computer & Information Science* (IEEE, New York, 2015)
17. C.-C. Chang , C.-J. Lin, LIBSVM: a library for support vector machines. ACM Trans. Intell. Syst. Technol. **2**(3), 27:1–27:27 (2011)
18. S. Pincus, Approximate entropy (ApEn) as a complexity measure. Chaos Interdiscipl. J. Nonlinear Sci. **5**(1), 110–C17 (1995)
19. L. Li, W. Chen, X. Shao, Z. Wang, Analysis of amplitude-integrated EEG in the newborn based on approximate entropy. IEEE Trans. Biomed. Eng. **57**(10), 2459–C66 (2010)
20. P.C. Petrantonakis, L.J. Hadjileontiadis, Adaptive emotional information retrieval from EEG signals in the time-frequency domain. IEEE Trans. Sig. Proc. **60**(5), 2604–2616 (2012)
21. F. Zhang, H. Meng, M. Li, Emotion extraction and recognition from music, in *International Conference on Natural Computation* (IEEE, New York, 2016)
22. Y. Zhu, S. Wang, Q. Ji, Emotion recognition from users' EEG signals with the help of stimulus VIDEOS, in *IEEE International Conference on Multimedia & Expo* (IEEE, New York, 2014)
23. O. Xie, Z.T. Liu, X.W. Ding, Electroencephalogram emotion recognition based on a stacking classification model, in *2018 37th Chinese Control Conference (CCC)* (2018), pp. 5544–5548

A Review of the Application of Deep Learning in the Classification of Diabetic Retinopathy

Xuyan Yu, Jianxia Liu, and Wenxuan Xue

Abstract Traditional image object classification algorithms and strategies are difficult to meet the requirements of image processing efficiency, performance, and intelligence. In recent years, deep learning in the computer vision has made great progress, showing good application prospects in medical image reading. Firstly, the background of deep learning and the knowledge of convolutional neural networks are introduced to fundamentally understand the basic model architecture and optimization methods of deep learning applied in the medical image. Secondly, the classification method of diabetic retinopathy images is discussed specifically. Finally, the problems faced in the future are analyzed.

Keywords Deep learning · Convolutional neural network · Algorithm model · Diabetic retinopathy

1 Introduction

Diabetic retinopathy (DR) is a serious blind eye disease. The long-term hyperglycemic environment can damage the endothelium of retinal blood vessels, causing a series of fundus lesions such as micro-angioma, hard exudation, cotton plaque, neovascularization, vitreous proliferation, and even retinal detachment [1]. However, the current detection of DR is a time-consuming manual process that requires an experienced clinician to examine and evaluate digital color fundus photographs of the retina. Doctors usually return the results of the test after a day or two. Such delays often increase the cost of communication between doctors and patients and miss the best treatment opportunity. In addition, some high-risk areas may not have enough ophthalmologists. So we need a comprehensive and automated DR screening method.

X. Yu · J. Liu (✉) · W. Xue
College of Information and Computer, Taiyuan University of Technology, Taiyuan, Shanxi, China

© Springer Nature Switzerland AG 2020
X. Yuan, M. Elhoseny (eds.), *Urban Intelligence and Applications*, Studies in Distributed Intelligence, https://doi.org/10.1007/978-3-030-45099-1_11

Feature expression is the key to processing medical images. For detecting DR, extracting lesion features is the most important link. Traditional feature design needs to be done manually, but this process is complex and requires high demands from the designer's technology. Therefore, the design of automated features has become an urgent need for efficient image processing [2–6].

Deep learning is an emerging field of machine learning research. Its core idea is to extract multi-level and multi-angle features from the original data through a series of nonlinear transformations in a data-driven manner so that the acquired features have greater generalization ability and expressive ability, which just meets the needs of efficient image processing [7]. The deep learning theory represented by the convolutional neural network has achieved breakthrough results in the field of image processing. This paper combines the basic principles of deep learning, focusing on the evolution and innovation of algorithms, models, and methods in the field of image processing.

2 Deep Learning

2.1 Background

Deep learning has a long and rich history. Neural networks were proposed in the 1950s. However, due to the lack of theoretical network training algorithms, insufficient training samples, and poor computing power of computers, neural network development encountered bottlenecks. As time goes on, the amount of available training data continues to increase; the advent of cloud computing and big data era has greatly improved the computer hardware infrastructure for training deep networks; the combination of unsupervised training strategy and the already proposed BP algorithm. Let the training of deep networks become possible. Therefore, deep learning has begun to be widely concerned.

2.2 Structure

The first area of application for deep learning is image recognition. The most important one is the convolutional neural network. The convolutional neural network is composed of three convolutional layers, pooled layers, and full-connected layers. Each layer has unique characteristics and functions.

- The convolution layer mainly extracts features, and the input feature map X and the K two-dimensional filters perform convolution operations and then outputs K two-dimensional feature maps. Convolution layer has two main points: (1) Local connection: Each neuron is only connected to a local area of the upper layer, and the space of the connection is called a local receptive field. (2) Weight sharing:

The current layer uses the same weight and bias for each channel's neurons in the depth direction.

Local connection and weight sharing reduce the number of parameters, greatly reducing training complexity and overfitting. At the same time, the convolution is robust to translation, rotation, and scale transformation on the image.

- The pooling layer compresses the input feature map, which makes the feature map smaller and simplifies the network computation complexity. On the other hand, feature compression is performed to extract the main features. At the same time, it also gives the model tolerance to mild deformation and improves the generalization ability of the model.

The full-connected layer is to connect all the features and send the output values to the classifier.

2.3   Neural Network-Based Optimization Method

As the number of neural network models becomes deeper and deeper, the training data becomes larger and larger, and the model structure becomes more and more complex. Network training often encounters problems such as overfitting, gradient disappearance, or gradient explosion. Here are a few neural network optimization methods:

- Hinton et al. proposed the "Dropout" [8] optimization technique, which is currently a common method to reduce the risk of overfitting in deep learning. It aims to inactivate a part of neurons randomly in the process of deep learning networks. This blocks the synergy between some neurons, reduces the joint adaptability between some neurons, makes the network have generalization ability, and reduces the risk of overfitting. Here is the workflow for Dropout:

  Suppose we want to train such a neural network, shown in Fig. 1, the input is x, the output is y, the normal flow is: we first forward x through the network forward, and then backpropagate the error to decide how to update the parameters for the network to learn.

Fig. 1 Standard neural network

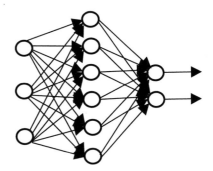

Fig. 2 Neural network after using dropout

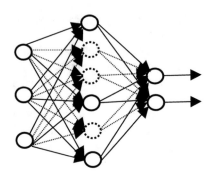

After using Dropout, the process is as follows:

a. Some hidden layer neurons are randomly deleted with a certain probability, shown in Fig. 2, and the input and output neurons remain unchanged (the dotted line in the figure is a partially deleted neuron)

b. The input x is forwarded through the modified network, and the resulting loss is propagated back through the modified network. After a small batch of training samples have been executed, the corresponding parameters (w, b) are updated according to the random gradient descent method on the neurons that have not been deleted.

- The essential principle of the Batch Normalization (BN) [9] is that when each layer of the network is input, a normalization layer is inserted, that is, a normalization process is performed first, followed by convolution processing. It is possible to fix the mean and variance of the input signals of each layer, effectively preventing the gradient explosion or disappearing.

If the output data of a certain layer A is normalized, this will affect the characteristics learned by the A-layer network. So the algorithm introduces the reconstruction of the learnable parameters γ, β, which is the key point of the algorithm.

$$y^{(k)} = \gamma^{(k)} \hat{\chi}^{(k)} + \beta^{(k)} \tag{1}$$

Each neuron x_k will have such a pair of parameters γ, β. Thus when

$$\gamma^{(k)} = \sqrt{\text{Var}\left[x^k\right]}, \beta^{(k)} = E\left[x^{(k)}\right] \tag{2}$$

can recover the features learned from a layer of the original layer. Therefore, the reconstruction parameters γ, β are introduced, so that our network can learn to recover the feature distribution that the original network has to learn. Finally, the forward conduction process formula of the Batch Normalization network layer is:

Input: values of x over a mini-batch: $\mathcal{B} = \{x_{1...m}\}$,

Parameters to be learned: γ, β

Output: $\{y_i = BN_{\gamma, \beta}(x_i)\}$

$$\mu_B \leftarrow \frac{1}{m} \sum_{i=1}^{m} x_i \tag{3}$$

$$\sigma_B^2 \leftarrow \frac{1}{m} \sum_{i=1}^{m} (x_i - \mu_B)^2 \tag{4}$$

$$\hat{x}_i \leftarrow \frac{x_i - \mu_B}{\sqrt{\sigma_B^2 + \epsilon}} \tag{5}$$

$$y_i \leftarrow \gamma \hat{x}_i + \beta \equiv BN_{\gamma, \beta}(x_i) \tag{6}$$

Equation (3) calculates the mini-batch mean; (4) calculates the mini-batch variance; (5) calculates the normalize; (5) calculates the scale and shift.

3 Classic Structure

3.1 AlexNet [10]

Deep learning technology was first applied to the direction of image recognition and achieved remarkable results. The AlexNet proposed by Alex et al. is the first deep convolutional neural network for image recognition, and the subsequent deep learning development in a series of image recognition is based on this. Compared to the traditional CNN structure, the AlexNet becomes deeper and wider. The network consists of five convolutional layers and three fully connected layers. The AlexNet establishes the dominant position of deep learning in image recognition, and also defines the general subject architecture of deep learning model in the field of image processing—feedforward convolutional neural network.

3.2 VGGNet [11]

The VGGNet is a deeper and broader evolution than the AlexNet. The VGGNet replaces a large convolution kernel with more small convolution kernels in series. This not only achieves the same convolution effect but also adds more nonlinear operations, enabling the network to extract more rich features while reducing the number of parameters. The VGGNet structure is shown in Table 1.

Table 1 VGGNet structure

ConvNet configuration					
A	A-LRN	B	C	D	E
11 weight Layers	11 weight Layers	13 weight Layers	16 weight Layers	16 weight Layers	19 weight Layers
Input (224 × 224 RGB image)					
conv3-64	conv3-64 LRN	conv3-64 conv3-64	conv3-64 conv3-64	conv3-64 conv3-64	conv3-64 conv3-64
Maxpool					
conv3-128	conv3-128	conv3-128 conv3-128	conv3-128 conv3-128	conv3-128 conv3-128	conv3-128 conv3-128
Maxpool					
conv3-256 conv3-256	conv3-256 conv3-256	conv3-256 conv3-256	conv3-256 conv3-256 conv3-256	conv3-256 conv3-256 conv3-256	conv3-256 conv3-256 conv3-256 conv3-256
Maxpool					
conv3-512 conv3-512	conv3-512 conv3-512	conv3-512 conv3-512	conv3-512 conv3-512 conv3-512	conv3-512 conv3-512 conv3-512	conv3-512 conv3-512 conv3-512 conv3-512
Maxpool					
conv3-512 conv3-512	conv3-512 conv3-512	conv3-512 conv3-512	conv3-512 conv3-512 conv3-512	conv3-512 conv3-512 conv3-512	conv3-512 conv3-512 conv3-512 conv3-512
Maxpool					
FC-4096					
FC-4096					
FC-1000					
Soft-max					

3.3 Google-Net

The Google Inception Net is known as the Inception V1 network. The new convolutional layer is also known as the Inception Module. The convolutional layer is divided into four parallel convolution operation, and the feature input of the upper layer is merged as an output through an aggregate operation after four lines of operation and then input to the next layer. The 1 × 1 convolution operation in the Module can not only organize information across channels, improve the expressive ability of the network, but also enhance and reduce the dimension of the output channel, simplifying the calculation process. The structure is shown in Fig. 3.

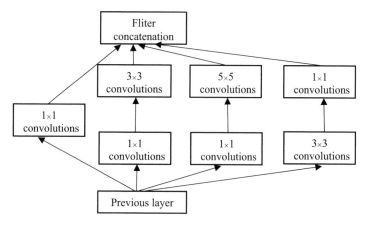

Fig. 3 Inception V1 structure

Fig. 4 Residual block

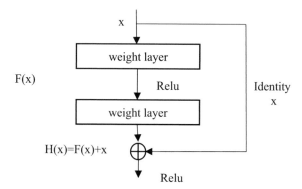

3.4 ResNet [12]

As the depth of the appropriate model continues to increase the number of layers, the accuracy of the model will decline. For this degradation phenomenon, He et al. proposed ResNet. The output of the Lth layer is no longer a single influence. The output of $L + 1$ also affects the output of the $L + 2$ layer, so every two layers can form a residual learning block, and the residual block changes the learning target in disguise. The entire ResNet is composed of multiple residual blocks, and the pool is composed of layers. The training process only needs to learn the difference between input and output, which protects information integrity and simplifies the learning objectives and difficulty. The residual block structure is shown in Fig. 4.

3.5 SE-Net [13]

Squeeze-and-Excitation (SENet) is the winner of the last ImageNet 2017 competition image classification. The SE block structure is shown in Fig. 5.

Given an input x whose channel feature number is c', a feature of the channel number c is obtained through a series of convolutions, such as the F_{tr} conversion operation in the framework:

$$F_{tr} : u_c = V_c \times X = \sum_{s=1}^{c'} v_c^s \times \chi^s \tag{7}$$

The first is the Squeeze operation, which performs feature compression along the spatial dimension. The Squeeze operation is actually a global average pooling:

Next is the Excitation operation:

$$s = F_{ex}(z, W) = \sigma(g(z, W)) = \sigma(W_2 \delta(W_1 z)) \tag{8}$$

Looking directly at the last equal sign, first multiply z by W_1, which is a fully connected layer operation. The dimension of W_1 is $C/r \times c$. This r is a scaling parameter, which is 16 in the design. The purpose of this parameter is to reduce the number of channels and thus reduce the amount of calculation. And because the dimension of z is $1 \times 1 \times c$, the dimension of $W_1 \times z$ is $1 \times 1 \times c/r$. After a ReLU layer, the output dimensions are unchanged. Then multiply by w_2, in fact, it is also a fully connected process, the dimension of w_2 is $c \times c/r$, so the output dimension is $1 \times 1 \times c$. Finally, after the sigmoid function, we get s whose dimension is $1 \times 1 \times c$. This s is the core of the entire framework. It is to describe the weight of c feature maps in U. After getting s, we can operate on the original U

$$\tilde{X}_c = F_{scale}(u_c, s_c) = s_c \cdot u_c \tag{9}$$

The SE block is not a complete network structure, but a substructure that can be embedded in other models of classification or detection.

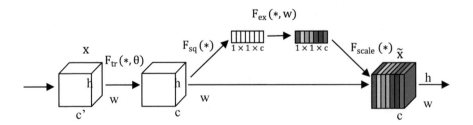

Fig. 5 A squeeze-and-excitation block

4 Diabetic Fundus Lesion Image Classification

4.1 Current Status of Fundus Image Classification Research

With the development of deep learning in medical image processing, there has been a lot of research on the classification of DR fundus images. Currently, strategies for classifying fundus images can be divided into two types. One is based on a local analysis strategy, that is, by detecting local lesions in the fundus image, one or more lesions are detected, and then the fundus images are classified or classified according to the type and number of lesions. The other is a global strategy, which is to analyze the overall characteristics of the entire fundus image and complete the classification and discrimination.

The DR automatic classification method is better and better from early simple classification based on information such as lesion structure and color, and later on more and more complex feature extraction and classification methods. Both classification methods based on local lesions and global images have their own characteristics. Most of the current methods are based on local lesions, have better clinical interpretation, and have sufficient characterization for each type of lesion, but require complex feature extraction methods. The global image classification method does not consider the structural characteristics of various types of lesions and avoids complicated feature extraction work.

4.2 Related Work

In the automatic analysis of fundus photographs, the existing work mainly focuses on the segmentation of blood vessels and optic discs. There is a method called Deep Vessel [14] that uses deep learning and conditional random fields for retinal vessel segmentation. The DRIU [15] approach proposes a unified framework for the segmentation of retinal vessels and optic discs. Since micro-angioma is usually used as a landmark lesion for DR discrimination, some methods [16] indirectly detect DR by detecting micro-angioma. The current classification criteria classify diabetic retinopathy into five categories, No DR, Mild DR, moderate DR, severe moderate DR, and proliferative DR.

4.3 Algorithm Flow

- Get image: At present, the largest open data source for DR detection is Kaggle diabetic retinopathy detection competition. The data sets all carry DR-type labels, category 0 corresponds to No DR, category 1 corresponds to Mild DR, and so on, and category 4 is proliferative.

- Image pre-processing: The background border needs to be removed; the image that is completely meaningless by noise pollution is deleted. For the third and fourth types of images to be dark, white balance and other methods are needed to improve the image darkness. It is also necessary to normalize the image. After normalization, all the data has been unified, which is very helpful in training the model. Since the experimental data set is extremely unbalanced, the type 0 data is nearly 36 times as large as the fourth type of data, so it is necessary to amplify the category with a smaller number of samples.

4.4 Model Design

For the CNN model design of DR detection, many of them are modified based on the above classical model to classify the data.

- The paper [17] uses the Google Inception V3 architecture as the base model. Then delete the top output layer of the base model and add two fully connected layers on the base model for feature passing. The first fully connected layer has 1024 units and the second fully connected layer has 512 units. The output layer uses a five-unit softmax classification layer. The credibility of the model is measured by accuracy, sensitivity, specificity, and kappa score. The structure is shown in Fig. 6 and Table 2.
- The article [18] designed the CompactNet neural network model, which is designed based on the AlexNet. The structure is shown in Fig. 7.

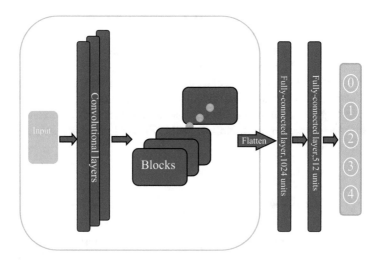

Fig. 6 Model structure

Table 2 Some indexes of the baseline model and this model

	Baseline [1]	Our model
Accuracy	73.76%	75.1540%
Sensitivity	95.05%	96.1500%
Specificity	29.99%	31.1423%
Kappa	0.4398	0.5107

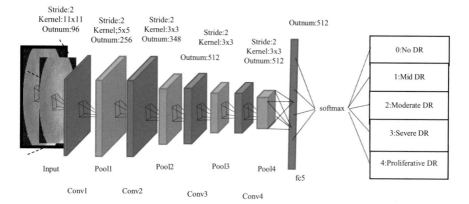

Fig. 7 CompactNet network architecture

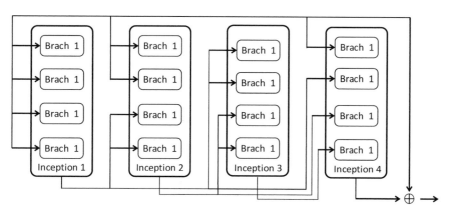

Fig. 8 Multi-branch design

And through experiments, the network is better than the traditional network training method, its classification index reaches 0.87, which is 0.27 higher than the traditional network.

- The article [19] is based on the Inception structure. By adding Residual links between Inception modules and modules, multi-branch links are built to achieve the corresponding design goals. The structure is shown in Fig. 8.

The results obtained using the regression feature were 0.812 with a kappa value of 0.833; the results obtained using the classification feature were 0.816 and the kappa value was 0.838.

5   Conclusion and Future Work

Deep learning is mainly based on convolutional neural network research in the image field, but the convolution operation is bound to have a large amount of computation in the whole network, which leads to network training taking a long time, changing the convolution operation form, and simplifying the computational complexity. It will also become a major development direction. At present, the small data set is a common problem surrounding medical image data sets. In fact, depending on factors such as cost and time, it is unrealistic to rely entirely on labeled data. Since there are more than 50,000 unlabeled data in the Kaggle competition, how to build a semi-supervised learning model could be an interesting research topic. Making full use of unmarked data is also an important development direction in the future.

References

1. H. Pratt, F. Coenen, D.M. Broadbent, et al., Convolutional neural networks for diabetic retinopathy. Proc. Comput. Sci. **90**, 200–205 (2016)
2. K. Shankar, M. Elhoseny, S.K. Lakshmanaprabu, M. Ilayaraja, R.M. Vidhyavathi, M. Alkhambashi, Optimal feature level fusion based ANFIS classifier for brainMRI image classification, in *Concurrency and Computation: Practice and Experience* (Wiley, Hoboken, 2018)
3. M. Elhoseny, K. Shankar, S.K. Lakshmanaprabu, A. Maseleno, N. Arunkumar, Hybrid optimization with cryptography encryption for medical image security in Internet of Things, in *Neural Computing and Applications* (Springer, New York, 2018). https://doi.org/10.1007/s00521-018-3801-x
4. K. Shankar, M. Elhoseny, R. Satheesh Kumar, S. K. Lakshmanaprabu, X. Yuan, Secret image sharing scheme with encrypted shadow images using optimal homomorphic encryption technique. J. Ambient Intell. Humanized Comput. 2018. https://doi.org/10.1007/s12652-018-1161-0
5. M. Elhoseny, G.-B. Bian, S.K. Lakshmanaprabu, K. Shankar, A.K. Singh, W. Wu, Effective features to classify ovarian cancer data in internet of medical things. Comput. Netw. **159**, 147–156 (2019)
6. E. Mohamed, K. Shankar, J. Uthayakumar, Intelligent diagnostic prediction and classification system for chronic kidney disease. Nat. Sci. Rep. **9**, 9583 (2019). https://doi.org/10.1038/s41598-019-46074-2
7. N. Krishnaraj, M. Elhoseny, M. Thenmozhi, M.M. Selim, K. Shankar, Deep learning model for real-time image compression in Internet of Underwater Things (IoUT). J. Real-Time Image Process. 2019. https://doi.org/10.1007/s11554-019-00879-6
8. N. Srivastava, G. Hinton, A. Krizhevsky, et al., Dropout: a simple way to prevent neural networks from overfitting. J. Mach. Learn. Res. **15**(1), 1929–1958 (2014)
9. S. Ioffe, C. Szegedy, Batch normalization: accelerating deep network training by reducing internal covariate shift. in *International Conference on International Conference on Machine Learning*. JMLR.org, 2015

10. A. Krizhevsky, I. Sutskever, G.E. Hinton, ImageNet classification with deep convolutional neural networks, in *International Conference on Neural Information Processing Systems* (Curran Associates Inc., Red Hook, 2012), pp. 1097–1105
11. K. Simonyan, A. Zisserman, Very deep convolutional networks for large-scale image recognition. arXiv preprint arXiv:1409.1556, 2014
12. K. He, X. Zhang, S. Ren, et al., Deep residual learning for image recognition. in *Proceedings of the IEEE Conference on Computer Vision and Pattern Recognition*, pp. 770–778, 2016
13. J. Hu, L. Shen, G. Sun, Squeeze-and-excitation networks. 2017
14. S. Xie, Z. Tu, Holistically-nested edge detection. Int. J. Comput. Vis. **125**(1-3), 3–18 (2015)
15. H. Fu, Y. Xu, D.W.K. Wong, et al., Retinal vessel segmentation via deep learning network and fully-connected conditional random fields, in *IEEE, International Symposium on Biomedical Imaging* (IEEE, Piscataway, 2016), pp. 698–701
16. X. Yuan, L. Gu, T. Chen, M. Elhoseny, W. Wang, A fast and accurate retina image verification method based on structure similarity. in *2018 IEEE Fourth International Conference on Big Data Computing Service and Applications*, pp. 181–185, 2018
17. L. Li, M. Fredrikson, S. Sen, et al., Case study: explaining diabetic retinopathy detection deep CNNs via integrated gradients. 2017
18. P. Ding, Q. Li, Z. Zhang, F. Li, Diabetic retinal image classification method based on deep neural network. J. Comput. Appl. **37**(3), 699–704 (2017)
19. H. Pang, C. Wang, Deep learning model for diabetic retinopathy detection. J. Softw. **28**(11), 3018–3029 (2017)

Part III
Smart Mobility and Transportation

Online Prediction Model of Short-Term Traffic Flow Based on Improved LS-SVM

Zhongjun Ma, Lin Feng, Zhenchun Wei, Zengwei Lyu, Zhensheng Huang, and Fei Liu

Abstract The change of urban road traffic flow has strong randomness and strong timeliness. The traditional short-term traffic flow prediction model has high complexity and strong closeness, and it is difficult to make full use of the previous learning results. In this paper, we analyze the support vector sparsity loss and parameter optimization time-consuming of Least Squares Support Vector Machine (LS-SVM) and propose a Dynamic Parameter Optimization LS-SVM named DPO-LSSVM for predicting short-term traffic flow. The experiments conducted on the real-world data set demonstrate the effectiveness of DPO-LSSVM on the training speed. Comparing with the Particle Swarm Optimization LS-SVM (PSO-LSSVM) and the RBF neural network online prediction model, the average training time of our model is only 30–60% of the other two.

Keywords Online prediction · Short-term prediction · Sparseness · Parameter optimization · LS-SVM

1 Introduction

In previous studies, short-term traffic flow predictions were mostly offline training models, while the actual prediction models were mostly online predictions. The important indicator is that the model parameters must be updated within a sampling time period. If the learning results of the previous step can be fully utilized, the calculation of the algorithm can be completed within one sampling time period [1].

Many offline prediction models are more complex and do not respond well to short-term predictions. Online learning algorithms are used as analytical tools for streaming data. The low computational complexity of algorithms is often associated

Z. Ma · L. Feng (✉) · Z. Wei · Z. Lyu · Z. Huang · F. Liu
School of Computer Science and Infomation Engineering, Hefei University of Technology, Hefei, Anhui, P. R. China
e-mail: fenglin@hfut.edu.cn; weizc@hfut.edu.cn; lvzengwei@mail.hfut.edu.cn; feiliu@mail.hfut.edu.cn

© Springer Nature Switzerland AG 2020
X. Yuan, M. Elhoseny (eds.), *Urban Intelligence and Applications*, Studies in Distributed Intelligence, https://doi.org/10.1007/978-3-030-45099-1_12

with low convergence rates [2]. Among them, the Least Squares Support Vector Machine (LS-SVM) has a faster operation speed [3, 4]. Because the LS-SVM model itself has the disadvantages of support vector sparsity loss and time-consuming parameter optimization process, this paper proposes a short-term traffic flow online prediction model based on an improved LS-SVM. This model aims to maintain a good predictive effect on the premise of ensuring stability.

2 Related Work

The learning algorithms of online prediction models currently have two directions: passive attack learning and adaptive regularization learning [5]. Xiang et al. [6] proposed a parameter optimization method based on uniform design to transform large sample searches into small sample search technology. Wang et al. [7] proposed a short-term prediction method based on PSO-SVM. Aiming at the problem of the loss of sparsity of the support vector machine model affects the efficiency of secondary learning. In [8, 9], the Sequence Minimum Optimization (SMO) algorithm and density-weighted sparse algorithm are used to sparse the solution.

Liu et al. [10] designed the incremental learning algorithm and an online learning algorithm of support vector machines. Wang et al. [11] proposed a traffic model that combines Adaptive Particle Swarm Optimization (APSO) and recursive least squares support vector machine regression. The online modeling and control method of the least squares support vector machine was introduced in detail in [12]. Wang et al. [13] proposed an online prediction research method based on Elman neural network. Reference [14] proposed an Online Nuclear Extreme Learning Machine (OL-KELM) method, which improved the online learning efficiency of the network. Ye et al. [15] presented a geographically weighted regression model for urban traffic black-spot analysis. Luo et al. [16] developed a dynamic taxi service planning method by minimizing cruising distance without passengers.

In this paper, a short-term traffic flow online prediction model PDO-LSSVM is proposed based on improved LS-SVM to accelerate the parameter optimization of LS-SVM. The prediction mode is expanded from offline to online so that the model has higher learning efficiency and better adapts to online prediction needs.

3 Online Prediction of Short-Term Traffic Flow on Urban Roads

3.1 Relevant Parameter

Denote the period number in the time series as t ($t \in \{1, 2, 3, \ldots, n\}$). The total sampling time is a certain period of 1 day, and the time series length can be calculated by

$$n = \frac{T_d}{T}, \tag{1}$$

where T_d is denoted as the total sampling time, T is the sampling statistical period (usually a fixed value), and N_t is denoted as the number of vehicles passing in the period t.

Traffic rate is introduced to measure the traffic flow, which satisfies Eq. (2).

$$V_{ti} = \frac{N_{ti}}{L_i \cdot N_{Li} / (l_c + l_d)}, \tag{2}$$

where V_{ti} is denoted as the traffic rate of road segment i between the period t, and N_{ti} represents the number of vehicles passing through the monitoring point in road segment i at the period t. Li is denoted as the length of road segment i, and N_{ti} is the number of lanes in road segment i. l_c and l_d are the standard vehicle length and a fixed safety distance between two cars, respectively.

3.2 Analysis of Problems

Compared with other machine learning algorithms, the model complexity of LS-SVM is lower. Denote the training set as $\{(x_1,y_1),(x_2,y_2), \ldots, (x_l,y_l)\}(x_i \in R^n, i = 1, 2, \ldots, l)$, where x_i contains the serial number of the current period and the historical traffic flow data of the current section and adjacent sections, y_i is the traffic flow data of the next period. The optimal decision function of the regression problem is as follows:

$$f(x) = \omega^T \cdot \Phi(x) + b, \tag{3}$$

where $\Phi(\cdot)$ represents a mapping the input space to a higher-dimensional space.

The evaluation target of the prediction problem is set to the Eq. (4). MAE is the mean absolute error. The value of l is related to the size of the test sample. x_i and y_i are the input and output of the test sample, respectively.

$$\min \frac{\sum_{i=1}^{l} |y_i - f(x_i)|}{l}. \tag{4}$$

The training samples of the traditional offline prediction model are fixed or updated in a long period, while the training samples of the online prediction model need to be updated in real-time according to the sampling period. On the basis of model prediction error, the average time of model training (denoted as T_{train}) was added to evaluate the effect of the online prediction model.

The average absolute error of the A and B prediction models is defined as E_A, E_B. The model that uses less time is better while $|E_A - E_B| / (E_A + E_B) \leqslant 1\%$, and the model with smaller error is a better while $|E_A - E_B| / (E_A + E_B) > 1\%$.

4   LS-SVM Prediction Model

4.1   Support Sparsity Loss of Vector

LS-SVM aims to convert Eq. (5) into Eq. (6) as follows.

$$
\begin{aligned}
\min \quad & J(\omega, \xi) = \left(\omega^T \cdot \omega\right)/2 + \gamma \sum_{i=1}^{l} \left(\xi_i + \xi_i^*\right) \\
\text{s.t.} \quad & y_i - \left(\omega^T \cdot \Phi(x_i) + b\right) \leq \varepsilon + \xi_i \\
& \omega^T \cdot \Phi(x_i) + b - y_i \leq \varepsilon + \xi_i^* \\
& \xi_i \cdot \xi_i^* \geq 0, i = 1, \ldots, l
\end{aligned}
\tag{5}
$$

$$
\begin{aligned}
\min \quad & J(\omega, e) = \left(\omega^T \cdot \omega\right)/2 + \gamma \left(\sum_{i=1}^{l} \xi_i^2\right)/2 \\
\text{s.t.} \quad & y_i = \omega^T \cdot \Phi(x_i) + b + \xi_i, \quad i = 1, \ldots, l
\end{aligned}
\tag{6}
$$

Although such error handling may bring us convenience, it causes sparseness of the solution, and all training samples become support vectors [10], support vector is the training sample point at the edge of the upper and lower intervals of the regression curve [17]. All the training samples are included in the support vector, so that the iterative optimization calculation amount increases during the model training and makes the prediction model susceptible to the abnormal point and leads to overfitting.

4.2   LS-SVM Parameter Optimization Algorithm

The core link of the support vector machine is the optimization of two parameters (c and g) in model training. c is the penalty coefficient, which affects the generalization ability of the model. g is the parameter of the RBF kernel function, which determines the distribution of the data mapping to the new feature space.

The variation range of the penalty parameter c is between $[2^{cmin}, 2^{cmax}]$, the variation range of the kernel parameter g is between $[2^{gmin}, 2^{gmax}]$, and the sizes of cstep and gstep are the optimal sizes of c and g (the default c and g are between $[2^{-8}, 2^{+8}]$, we set cstep and gstepas 1). The relatively optimal combination of c and g is selected by calculation error.

We select the combination of c and g to train the LS-SVM and calculate the model prediction error, the optimization target is the minimum error of the intelligent algorithm.

5   Improving the LS-SVM

But LS-SVM only considers the equality constraint and loses the sparsity of the support vector machine solution, which affects the efficiency of secondary learning. The loss of sparsity of the solution and the time-consuming of the parameters optimization have little effect on the prediction. However, increase in sample size leads to an increase in abnormal points, which is not applicable to the online prediction model.

5.1 Support Vector Sparse Strategy

At present, we always use two kinds of methods to construct the sparsity of the solution process: clustering method to eliminate outliers; SMO algorithm to pruning the solution. But high-dimensional clustering requires dimensionality reduction on the sample first, the increase in computational complexity leads to serious time-consuming, and the SMO algorithm is relatively simple and efficient to operate. For the short-term traffic flow prediction problem of urban roads, the LS-SVM supports the vector sparse SMO strategy as shown in Algorithm 1.

5.2 Parameter Optimization Improvement Strategy

We propose a parameter optimization strategy based on the fixed-parameter optimization range and optimization step size of the traditional Grid-search algorithm, and propose an improved strategy for parameter optimization, as shown in Algorithm 2.

Algorithm 1 Support vector sparse strategy for LS-SVM

Step (1) Perform preliminary parameter optimization on the training samples to obtain the current optimal parameter combination

Step (2) Calculate the error corresponding to each training sample under the current parameter combination

Step (3) Grouped by time series, eliminate the sample with the smallest absolute value of each group, and update the training samples

Step (4) Stop if the magnitude of the model support vector is less than the threshold, otherwise, return to step (1)

Algorithm 2 Improved Grid-search parameter optimization strategy

Step (1) *Initialize*: the boundaries of parameters c and g: $c_{minedge}$, $c_{maxedge}$, $g_{minedge}$, $g_{maxedge}$, and the step

Step (2) Use the Grid-search method to perform a grid search in the boundary range and return the optimization result: (c_{opt}, g_{opt})

Step (3) Update the boundary range and the optimization according to Eq. (7) (Increase the boundary value. Note, the step takes integer numbers.)

Step (4) *Loop* 2, 3 steps *until* step is the minimum threshold or the number of cycles is the maximum number of recursive iterations

Step (5) *Return* (c_{opt}, g_{opt})

$$c_{\min} = c_{opt} - 2step$$
$$c_{\max} = c_{opt} + 2step$$
$$g_{\min} = g_{opt} - 2step \qquad (7)$$
$$g_{\max} = g_{opt} + 2step$$
$$step = \tfrac{1}{2}step$$

The grid number of Grid-search optimization grid is used by us to analyze the optimization strategy before and after improvement. The number of grid blocks indirectly represents the maximum number of decision trees. After each initial refinement, the length of each recursive iteration c and g is $4n$, and the optimal step size is $n/2$, so the number of optimized grid blocks is 64 and the maximum number of recursions is 1+. Improving the Grid-search optimization strategy and the default time ratio as the following:

$$\frac{\frac{x^2}{n^2} + 64 \cdot \log_2 n}{x^2} = \frac{1}{n^2} + \frac{64 \cdot \log_2 n}{x^2} \qquad (8)$$

The value of the optimal step n is related to the distribution of the combination of the model c and g. When n is close to the maximum step, we pick the largest number of parameter combinations for the operation.

5.3 Improving the Online Prediction Model of LS-SVM

As the repeated calculation of the model increases, we propose a comprehensive improvement strategy called DPO-LSSVM to adapt to the online prediction of LS-SVM. The detailed steps of the improved strategy are shown in Algorithm 3.

Algorithm 3 DPO-LSSVM online prediction model

Step (1) Initialize the boundaries of parameters c and g: $c_{minedge}$, $c_{maxedge}$, $g_{minedge}$, $g_{maxedge}$, and the step

Step (2) If it is not the first parameter optimization of the model, the previous optimization result (c_{opt}, g_{opt}) is used and step (4) is performed; otherwise, step (3) is performed

Step (3) The grid search is performed by the Grid-search method in the current parameter range and the optimal step size, and the optimization result (c_{opt}, g_{opt}) is updated

Step (4) If the magnitude of the support vector is greater than or equal to the threshold, step (5) is performed; otherwise, step (6) is performed

Step (5) Calculate the error corresponding to each training sample under the current parameter combination, group by time series number, eliminate the sample with the smallest absolute value of each group of errors, and then organize the training samples, return to step (4)

Step (6) Update the boundary range and the optimization step strategy according to Eq. (7)

Step (7) If step is the minimum threshold or the number of loops is the maximum number of recursive iterations, step (8) is performed; otherwise return to step (3)

Step (8) Take the latest optimization results (c_{opt}, g_{opt}) to train the model and predict the traffic flow for the corresponding time period

Step (9) If there is no new training sample data, stop the loop; otherwise add a new training sample, and then eliminate the samples of the earliest same time period, and return to step (1)

---

6   Simulation

6.1   Data Set

Get traffic data from Dublin's official data portal (data.gov.ie). Three intersections recorded as C1, C2, and C3 for experiments were selected by us. The peak traffic flow of C1 was small, the peak traffic flow of C2 was middle, and the peak of the traffic flow of C3 was higher. This paper used the data acquired on the working days from February 6–10, 2012. The sampling period of the monitoring point is 6 min, 240 samples/day for a single monitoring point. The number of total sample capacity for 5 days is 3600.

The input and output matrix of the model is shown in Table 1, where $x_1 \sim x_9$ are model inputs, x_{10} is the forecast result output, and the space-time dependent attribute is expressed in the space-level connection and the time interval. In the calculation of the error, the flow rate needs to be converted into the flow to calculate.

The model training set we use is fixed to 80% of all samples (the sample capacity of 4 days), and the sample update period is set to 30 min, so the simulation experiment performs 48 model updates. Then we update training samples and train the model, and then use the trained model to predict for short-term traffic flow for the next 5 time periods and predict traffic flow for 1 day completely after 48 cycles. Finally, we calculate the average prediction error and the average training time of the model. The calculation method of the model prediction error is fixed, and the measurement of the model training time is relative.

Table 1   Descriptions of the input and output matrix parameter

Symbol	Parameter meaning
x_1	Time series
x_2	Flow rate of the previous period of the direct link
x_3	Flow rate of the current period of the direct link
x_4	Flow rate of the previous period of the right link
x_5	Flow rate of the current period of the right link
x_6	Flow rate of the previous period of the left link
x_7	Flow rate of the current period of the left link
x_8	Predict the flow rate of the road in the previous time period
x_9	Predict the flow rate of the road in the current time period
x_{10}	Predict the flow rate of the road in the future time period

6.2   Comparative Experiment

6.2.1   DPO-LSSVM

The c and g ranges of the LS-SVM model are set to $[10^{-1}, 10^{3}]$, we improve the search range and the optimal step size dynamics of the Grid-search algorithm, and the kernel function selects the RBF kernel with higher efficiency for solving the nonlinear regression problem.

Due to the dynamic adjustment optimization range and the optimization step size, there is a certain probability to fall into the local optimal solution compared with the traditional method. For the first model training, four optimization step sizes were selected, and four prime numbers 5, 7, 11, and 19 were taken. The optimal step size corresponding to the optimal parameter combination was selected as the initial optimization step size of the experiment, and the experimental data set was selected as 11.

The training samples were updated every 30 min and the model was trained using an improved strategy, cycled 48 times. The comparison between the predicted traffic flow and the real traffic flow at the C2 point of DPO-LSSVM is shown in Fig. 1.

6.2.2   Particle Swarm Optimization LS-SVM

Here, the velocity update formula of the i-th particle of the particle swarm optimization algorithm in the d-dimensional search is as shown in the Eq. (9), where w_v is the inertia weight to adjust the search range of the solution space, c_1 and c_2 search acceleration constants to adjust the maximum step size (local and global acceleration constant) of the search; we also introduce two random numbers r_1 and

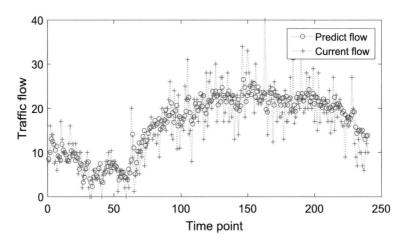

Fig. 1 Online prediction of short-term traffic flow at the C2 point of the DPO-LSSVM

r_2 to increase the randomness of the search. Equation (9) represents the weighted sum of the three parts of the i-th particle's previous update rate, the difference from the optimal solution, and the difference from the optimal solution between the populations.

$$v_{id}^k = w_v v_{id}^{k-1} + c_1 r_1 \left(\text{pbest}_{id} - x_{id}^{k-1} \right) + c_2 r_2 \left(\text{gbest}_d - x_{id}^{k-1} \right)$$
$$s.t. \ c_1, c_2 \in [0, 1] \tag{9}$$
$$w_v > 0$$

The position update formula of the i-th particle of the particle swarm optimization algorithm in the d-dimensional search is as shown in the Eq. (10), where wp denotes the inertia weight to adjust the search speed of the solution space.

$$x_{id}^k = x_{id}^{k-1} + w_p v_{id}^{k-1} \tag{10}$$

The ranges of c and g of the LS-SVM are set to $[10^{-1}, 10^3]$, and the relevant parameters of the particle swarm algorithm are set as follows: $i_{max} = 20$, $d_{max} = 200$, $c_1 = 1$, $c_2 = 1.7$, $w_v = 1$, $w_p = 1$; here, we set the fitness function of the particles in the population to the predicted mean square error of the validation set, as shown in Eq. (11). Because the particle swarm optimization algorithm is used to predict the actual sample and the error cannot be zero, the particle swarm optimization algorithm is terminated by reaching the maximum number of iterations. The training samples are updated every 30 min. First, we use the particle swarm optimization algorithm for optimizing the parameters, and then the model is retrained and cycled 48 times.

$$\frac{\sum_{i=1}^{l} (y_i - f(x_i))^2}{l} \tag{11}$$

The fitness curve of the Particle Swarm Optimization LS-SVM (PSO-LSSVM) in the first round is shown in Fig. 2. The fitness of more than 80% of the models is stable before the 100th generation, and the global average convergence is in the 77th generation.

The comparison between the predicted traffic flow and the real traffic flow at the C2 point with the PSO-LSSVM is shown in Fig. 3.

6.2.3 Neural Network Prediction Model

Here, we set the artificial neural network (ANN) with a mean square error target of 0.001, a learning rate of 0.1, a loop iteration number of 500 (control training duration, enough to converge), and the number of hidden nodes using the empirical Eq. (12), where m is a hidden layer node, n is an input layer node, α is a constant between 1 and 10, and the final number of hidden nodes is set to 5. In order to

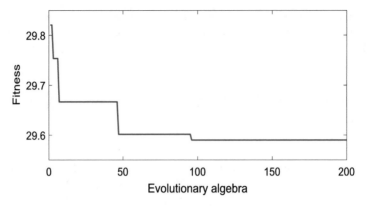

Fig. 2 The first round of fitness curve change chart under PSO-LSSVM

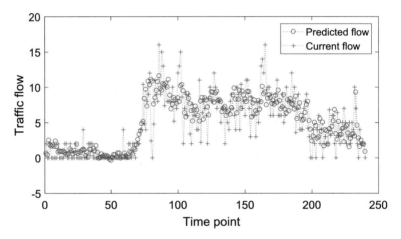

Fig. 3 Online prediction of short-term traffic flow at C2 point under PSO-LSSVM

provide nonlinear variation, the hidden layer selects the double S function tanh as the transfer function, the output layer selects the S function sigmoid, and the training function of the model selects the quasi-Newton algorithm training.

$$m = \sqrt{n+1} + \alpha \tag{12}$$

The training samples are updated every 30 min, then retrained and cycled 48 times. The comparison between the predicted traffic flow and the real traffic flow at the C2 point with the ANN model is shown in Fig. 4.

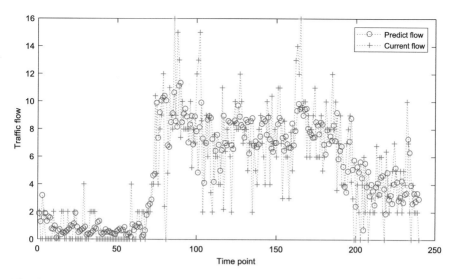

Fig. 4 Online prediction of short-term traffic flow at the C2 point of the ANN model

Table 2 Comparison of experimental errors in short-term traffic flow online prediction model

Prediction error	Points	C1	C2	C3
	Algorithm			
MAE	DPO-LSSVM	1.9918	4.1599	6.2503
	PSO-LSSVM	1.9965	4.1833	6.5032
	ANN	2.0083	4.0586	6.7939
MSE	DPO-LSSVM	6.9534	24.6133	71.7571
	PSO-LSSVM	6.9260	25.0829	75.7672
	ANN	7.1823	24.6299	84.0026

7 Summary

The experimental error pairs of the three short-term traffic flow online prediction models are shown in Table 2. The difference between the error of the ANN and the other two models exceeds 1%, and the difference ratio of the prediction errors of the three models exceeds 1%. The mathematical methods used in the support vector machine model are similar, so the error between the DPO-LSSVM and the PSO-LSSVM is closer.

Table 3 shows the average training time (s) of the three online prediction models, representing the training time of the single model, where CPUTime is the duration of the model training CPU, and CodeTime is the execution time of the program code.

The three online prediction models show their prediction errors in each round of the model as shown in Fig. 5. The minimum error times of DPO-LSSVM, PSO-LSSVM, and ANN are 20, 11, and 17, respectively. The smallest average prediction error is DPO-LSSVM.

Table 3 Comparison of average training time consumption of short-term traffic flow online prediction model

Time-consuming Algorithm	CPUTime	CodeTime
DPO-LSSVM	188.9195	0.0060
PSO-LSSVM	592.7206	0.1732
ANN	312.0215	2.5015

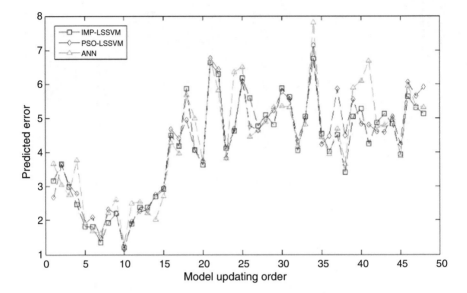

Fig. 5 Prediction error of each round of the online prediction model

Referring to the evaluation criteria proposed above, the DPO-LSSVM is optimal at the C1 and C3 monitoring points, and the ANN model is optimal at the C2 monitoring point. The DPO-LSSVM is optimal according to the average prediction error and training time. Although the average prediction error is improved by less than 5%, the average training time of improved LS-SVM is only 30% of the Particle Swarm Optimization LS-SVM.

Acknowledgments This work is supported by the National Key Research Development Program of China (2016YFC0801804), National Natural Science Foundation of China (61701162) and Fundamental Research Funds for the Central Universities of China (PA2019GDPK0079).

References

1. B. Hong, X. Jiang, Y. Qi, Q. Cheng, Research on LSSVM online prediction algorithm based on inverse of Hermite matrix. J. Syst. Simul. **29**(1), 1–6 (2017)
2. Z. Li, Y. Li, F. Wang, A survey of online learning algorithms for big data analysis. J. Comput. Res. Dev. **52**(8), 1707–1721 (2015)

3. J.A.K. Suykens, J. Vandewalle, B.D. Moor, Optimal control by least squares support vector machines. Neural Netw. **14**, 23–35 (2001)
4. J.A.K. Suykens, J.D. Brabanter, L. Lukas, et al., Weighted least squares support vector machines: robustness and sparse approximation. Neurocompting **48**(1), 85–105 (2002)
5. Z. Pan, S. Tang, J. Qiu, H. Guyu, Overview of online learning algorithms. J. Data Acquis. Process. **31**(6), 1067–1082 (2016)
6. C. Xiang, Z. Zhou, L. Zhang, Improved algorithm of least squares support vector machine based on uniform design. Comput. Simul. **28**(3), 194–197 (2011)
7. Y. Wang, D. Li, C. Gao, H. Zhang, Short-term load forecasting based on improved PSO-SVM. Elect. Meas. Instrum. **52**(3), 22–25 (2015)
8. Y. Liu, Y. Sheng, C. Jiang, L. Chen, Short-term load forecasting based on LS-SVM and SMO sparse algorithm. Power Syst. Protect. Contr. **36**(4), 63–66 (2008)
9. G. Si, H. Cao, Y. Zhang, L. Jia, A least squares support vector machine sparse algorithm based on density weighting. J. Xi'an Jiaotong Univ. **43**(10), 11–15 (2009)
10. N. Xiong, B. Liu, On-line prediction of network traffic based on adaptive particle swarm optimization LSSVM. Comput. Appl. Softw. **9**, 21–24 (2013)
11. W. Wang, C. Men, W. Lu, Online prediction model based on support vector machine. Neurocomputing **71**(4-6), 550–558 (2008)
12. X. Zhou, *Research on Online Modeling and Control Method Based on Least Squares Support Vector Machine* (Hunan University, Changsha, 2012)
13. Q. Wang, Q. Liu, On-line prediction of sinter quality based on Elman neural network. Instrum. Tech. Sens. **2017**, 10 (2017)
14. C. Ma, Y. Zhang, Z. Li, G. Yi, On-line prediction of characteristic parameters of hydraulic pump based on nuclear extreme learning machine. Comput. Simul. **5**, 351–354 (2014)
15. Y. Ye, Z. Zuo, X. Yuan, S. Zhang, X. Zeng, Y. An, B. Chen, Geographically weighted regression model for urban traffic black-spot analysis, in *2017 IEEE International Geoscience and Remote Sensing Symposium*
16. Z. Luo, H. Lv, F. Fang, Y. Zhao, Y. Liu, X. Xiang, X. Yuan, Dynamic taxi service planning by minimizing cruising distance without passengers. IEEE Access. **6**, 70005–70016 (2018)
17. B Giritharan, S Panchakarla, X Yuan, Segmentation of CE videos by finding convex skin. in *2010 IEEE International Conference on Bioinformatics and Biomedicine Workshops (BIBMW)*, 2010. pp. 158–163

Research on Low Altitude Object Detection Based on Deep Convolution Neural Network

Yongjun Qi, Junhua Gu, Zepei Tian, Dengchao Feng, and Yingru Su

Abstract The rapid and accurate detection of low altitude objects means a great deal to flight safety in low altitude airspace; however, low altitude object detection is very challenging due to the images' characteristics such as scale variations, arbitrary orientations, extremely large aspect ratio, and so on. In recent years, deep learning methods, which have demonstrated remarkable success for supervised learning tasks, are widely applied to the field of computer vision and good results have been achieved. Therefore, the deep learning method is applied to low altitude object detection in this paper. We proposed a deep convolution neural network model, which utilizes deep supervision implicitly through the dense layer-wise connections and combines multi-level and multi-scale feature. The model has achieved state-of-the-art performance on two large-scale publicly available datasets for object detection in aerial images.

Keywords Low altitude safety · Object detection · Deep learning

Y. Qi · J. Gu
State Key Laboratory of Reliability and Intelligence of Electrical Equipment, Hebei University of Technology, Tianjin, China

Information Technology Center, North China Institute of Aerospace Engineering, Langfang, China

J. Gu (✉)
State Key Laboratory of Reliability and Intelligence of Electrical Equipment, Hebei University of Technology, Tianjin, China

Hebei Province Key Laboratory of Big Data Calculation, Hebei University of Technology, Tianjin, China

Z. Tian
School of Artificial Intelligence, Hebei University of Technology, Tianjin, China

D. Feng . Y. Su
Information Technology Center, North China Institute of Aerospace Engineering, Langfang, China

© Springer Nature Switzerland AG 2020
X. Yuan, M. Elhoseny (eds.), *Urban Intelligence and Applications*, Studies in Distributed Intelligence, https://doi.org/10.1007/978-3-030-45099-1_13

169

1 Introduction

Low altitude airspace is an important strategic resource for all countries in the world [1]. In recent years, some major developed countries and China have successively opened some pilot areas of low altitude airspace. Thanks to the lack of uniform industry standards and norms, non-cooperative intrusion flights of low altitude airspace have become commonplace at home and abroad, which endanger the privacy of citizens, the safety of life and property, but also pose a great threat to public security. Therefore, low altitude objects detection is great significance to ensure the safety of low altitude flight. However, low altitude object detection is very challenging due to the following characteristics of low altitude object images [2].

- The scale variations of object instances in aerial images are huge. This is not only because of the spatial resolutions of sensors, but also due to the size variations inside the same object category.
- Many small object instances are crowded in aerial images. Moreover, the frequencies of instances in aerial images are unbalanced, some small-size images contain many instances, while some large-size images may contain only a handful of small instances.
- Objects in aerial images often appear in arbitrary orientations. In addition, there are also some instances with an extremely large aspect ratio, such as a bridge.

Some literatures [3–6] attempt to transfer object detection algorithms developed for natural scenes to the aerial image domain. There are other researchers who have pursued approaches based on fine-tuning networks pre-trained on large-scale image datasets, such as ImageNet [7] and MSCOCO [8], for detection in the aerial domain.

Unlike the above methods, we propose a deep convolution neural network model, which utilizes deep supervision implicitly through the dense layer-wise connections proposed in DenseNet [9]. The model can train object detection networks from scratch and identify diverse objects on low altitude remote sensing datasets. Our model has achieved state-of-the-art performance on two large-scale datasets for object detection in aerial images.

2 Related Work

Object detection has been actively studied for the last few decades [10, 11]. Traditional object detection methods are built on handcrafted features and shallow trainable architectures. The pipeline can be mainly divided into three stages: informative region selection, feature extraction, and classification. However, traditional object detection methods have the following problems.

- The generation of candidate bounding boxes with a sliding window strategy is redundant, inefficient, and inaccurate.
- Feature extraction relies heavily on prior knowledge.

- The semantic gap cannot be bridged by the combination of manually engineered low-level descriptors and discriminatively trained shallow models.

The availability of large scales training data, such as ImageNet [7], the raise of high-performance computing systems, such as GPUs, and various CNN (Convolutional Neural Network) based methods [12] have been substantially improving upon the performance of object detection. The new era of object detection is opened by deep convolutional neural network models, such as VGG [13], GoogLeNet [14], ResNet [15], etc., which contain thousands of parameters and have the ability to extract rich features and context information. Many end-to-end object detection frameworks are derived based on these deep convolution neural network components. They can be classified into two categories: region-proposal based framework such as Fast R-CNN [16], Faster-RCNN [17], FPN [18], and regression-based framework, such as YOLO [19], SSD [20]. In comparison, the region-proposal based framework has better detection accuracy, and the regression-based framework has a faster detection speed.

In recent years, deep convolutional neural networks [21–24] have been widely used in low altitude object detection. In order to deal with the problem of aircraft detection, Cheng et al. [3] proposed a method to learn a rotation-invariant and fisher discriminative CNN model, which adopted the AlexNet [12] pre-trained on ImageNet. Long et al. [4] proposed an object localization framework based on CNN for tackling the problem of automatic accurate localization of aircraft objects in high-resolution remote sensing images. Zhu et al. [25] proposed a novel framework with two deep learning components including FCN [26] and deep CNN for detecting and recognizing traffic signs. Aiming at automatically recognizing birds at small scale together with large background regions, Takekl et al. [27] proposed a framework based on CNN and FCN. Wang et al. [28] proposed a multi-channel deep neural network model for small unmanned aerial vehicle object recognition tasks. Lu et al. [29] employed the hierarchical feature layers from CNN for object detection and recognition. Yuan et al. [30] proposed a regularized ensemble framework of deep learning for target detection from multi-class, imbalanced training data. Yuan and Sarma [31] leveraged ALSM data for automatic urban water-body detection and segmentation based on spatially constrained model-driven clustering.

The proposed model is different from those that rely heavily on the off-the-shelf networks pre-trained on large-scale classification datasets like ImageNet in the paper. It can train object detectors from scratch, and detect multiple instances at a time.

3   Proposed Method

In this section, we introduce the network model in detail. Our model can be seen as two parts; the first part is feature extraction using a dense block that originates from DenseNet. The second part is inference based on multi-level and multi-scale feature maps. Figure 1 illustrates the model structure.

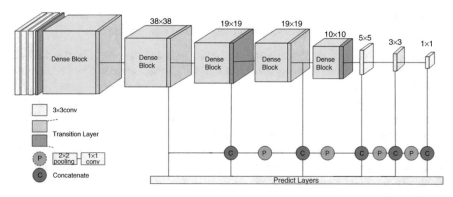

Fig. 1 Network model. The model consists of convolution layers, pooling layers, dense blocks, and transition layers. Dense block includes six 1×1 and 3×3 combined convolution layers. Transition layer has a 1×1 convolution layer and a 2×2 pooling layer (pink) or only one 1×1 convolution layer (purple). Multi-scale prediction consists of feature pyramids with six scales (38 $\times$ 38, 19 $\times$ 19, 10 $\times$ 10, 5 $\times$ 5, 3 $\times$ 3, 1 $\times$ 1)

3.1 Dense Block

Inspired by the DenseNet [9], we define dense block as a composite function of four consecutive operations: convolution (Conv), followed by batch normalization (BN) and a rectified linear unit (ReLU), scale layer (Scale). The growth rates [9] of dense block 1–5 are 48. Inside the blocks, each layer receives all the preceding layers' feature maps as inputs. The structure can enhance feature propagation and support feature reuse. It is helpful for tasks with less obvious features.

3.2 Transition Layer

The layers between two adjacent dense blocks are referred to as transition layers including a convolution layer and a pooling layer. The number of feature maps inside each dense block increases dramatically. $1 \times 1 \times 1$ convolution layers can compress the information channel-wise and reduces the total number of channels throughout the network. However, experiments show that the performance is better without the pooling layer in some transition layers. It may be due to the fact that pooling reduces the resolution of feature maps and loses too much information. If each layer contains a pooling layer, the resolution will become smaller and smaller.

3.3   Multi-Level and Multi-Scale Feature

Inspired by the SSD [20] model, we add a multi-scale feature map prediction. Each later scale is directly transited from the adjacent previous scale. Compared with SSD, we choose a more complex but effective structure, which fuses multi-scale information for each layer. In each scale (except scale 1), half of the feature maps are learned from the previous scale with a series of conv-layers. Shallow feature maps learn the regional features while deep feature maps learn the abstract features. We analyze that abstract features have rich semantics, which is helpful to improve the performance of object recognition, while regional features are sensitive to the location and size of objects, which is helpful to improve the performance of object detection. More accurate results can be obtained by combining shallow and deep features.

4   Experiments

In this section, a description of the datasets is first provided. After that, we have designed three groups of experiments. The first group of experiments aims to verify the optimal value of key hyperparameter—growth rate and the optimal feature fusion method. Based on the optimal setup, the other two groups of experiments are carried out on two large-scale datasets, respectively. Object detection performance was measured by mean Average Precision (mAP). The algorithm is implemented under the open-source Caffe framework. The server configurations are Intel Xeon E5-2683v4 CPU, 128GB RAM, and two NVIDIA Quadro P5000.

4.1   Datasets

We trained and evaluated our model on two publicly available datasets: NWPU VHR-10 [32] and DOTA [2]. NWPU VHR-10 contains 10 categories and a total of 800 very high-resolution optical remote sensing images, where 715 color images were acquired from Google Earth with the spatial resolution ranging from 0.5 to 2 m, and 85 pansharpened color infrared images were acquired from Vaihingen data with a spatial resolution of 0.08 m.

DOTA contains 2806 aerial images from different sensors and platforms. Each image is of the size of about 4000 × 4000 pixels and contains objects exhibiting a wide variety of scales, orientations, and shapes. These images are then annotated by experts in aerial image interpretation using 15 common object categories. The fully annotated images contain 188,282 instances, each of which is labeled by an arbitrary quadrilateral.

4.2   Experimental Process and Results

Growth Rate Growth rate can control the output channel of each convolution layer in the block. Assume the number of channels in the input layer of a block is $c0$ and the block has i convolution layers with a growth rate of g. Then the output of the block will have $c0 + i \times g$ channels. In the experiment, the growth rate was set to 16, 32, and 48, respectively.

Feature Fusion We propose the feature fusion structure for prediction. Three feature fusion methods, weight sum, weight product, and channel concatenation, are verified in experiments.

The experimental results of different parameter setup on the NWPU VHR-10 are shown in Table 1.

We can see clearly from Table 1 that the performance is the best when the growth rate is set to 48 and the mAP is 5.13 and 1.21 higher than 16 and 32, respectively. In the same way, the performance of Concatenation is the best among the three feature fusion methods and the mAP is 1.06 and 1.73 higher than Sum and Product, respectively.

Based on optimal hyperparameter, we run experiments on the widely used NWPU VHR-10 and DOTA datasets. To quantitatively evaluate the proposed model, we compared it with four state-of-the-art methods, in which, the spatial sparse coding BoW (SSCBoW) feature [33], the sparse-coding-based feature (FDDL) [34], the collection of part detectors (COPD) [32], and the single shot multibox detector (SSD) [20] are employed, respectively. Table 2 shows the quantitative comparison results on NWPU VHR-10.

Figure 2 shows object detection results on NWPU VHR-10 with the proposed approach.

Table 1 The results of different parameter setup ON THE NWPU VHR-10 (MAP, %)

Categories	Growth rate			Feature fusion		
	16	32	48	Sum	Product	Concatenation
Airplane	63.33	74.96	73.12	77.03	75.22	76.03
Ship	60.25	71.89	69.25	72.52	73.85	71.57
Storage tank	52.42	53.33	**55.16**	77.83	75.95	61.46
Baseball diamond	86.97	85.9	85.47	86.51	89.11	87.34
Tennis court	65.86	66.93	**71.32**	66.04	68.65	**70.20**
Basketball court	77.79	79.12	**83.77**	82.68	80.13	**83.64**
Ground track field	87.21	86.04	84.65	85.83	89.56	83.93
Harbor	62.42	62.71	60.13	65.23	67.23	59.68
Bridge	65.51	70.1	**70.37**	62.27	47.95	**74.39**
Vehicle	54.66	63.63	**74.84**	54.55	56.11	**72.84**
Mean AP	67.64	71.56	**72.77**	73.05	72.38	**74.11**

The best results are highlighted in bold face

Table 2 Object detection results on NWPU VHR-10 (mAP, %)

Categories	Methods				
	SSCBoW	FDDL	COPD	SSD	Our method
Airplane	50.61	29.15	62.25	84.4	74.96
Ship	50.84	37.64	68.87	79.12	72.89
Storage tank	33.37	77.00	63.71	44.19	**53.33**
Baseball diamond	43.49	25.76	83.27	98.13	85.9
Tennis court	0.33	2.75	32.08	58.6	**66.93**
Basketball court	14.96	3.58	36.25	71.32	**79.12**
Ground track field	10.07	20.10	85.31	39.42	**86.04**
Harbor	58.33	25.39	55.27	32.21	**62.71**
Bridge	12.49	21.54	14.79	56.12	**70.1**
Vehicle	33.61	4.47	44.03	41.31	**63.63**
Mean AP	30.81	24.74	54.58	60.48	**71.56**

The best results are highlighted in bold face

Fig. 2 Detection results on NWPU VHR-10 with the proposed approach

In contrast to NWPU VHR-10, DOTA has more kinds of categories and instances, and it is more difficult for object detection. On DOTA, we only compare the best performing methods (SSD) of four methods with our method in Tables 1. In addition, both of which belong to the regression-based framework. Tables 3 shows the quantitative comparison results on DOTA.

Figure 3 shows a number of object detection results with the proposed approach, despite the large variations in the orientations and sizes of objects, the proposed approach has successfully detected and located most of the objects.

From the above experimental results, we can see that our model shows good performance in both NWPU VHR-10 and DOTA. The main reason may be that Dense block can fuse multiple layers of image features, which makes the feature

Table 3 Object detection results on DOTA (map, %)

Categories	Methods	
	SSD	Our method
Plane	44.74	**48.58**
Baseball diamond	11.2	**16.94**
Bridge	6.22	**9.21**
Ground field track	6.91	**25.36**
Small vehicle	2	**20.2**
Large vehicle	10.24	**44.49**
Ship	11.34	**12.51**
Tennis court	15.59	**75.36**
Basketball court	12.56	**15.76**
Storage tank	17.94	**19.09**
Soccer-ball field	14.73	**20.21**
Roundabout	4.55	**19.09**
Harbor	4.55	**12.65**
Swimming pool	0.53	**16.41**
Helicopter	1.01	**4.54**
Mean AP	10.94	**24.03**

The best results are highlighted in bold face

Small vehicle   Large vehicle   Plane

Fig. 3 Detection results on DOTA with the proposed approach

map richer and more capable of capturing small targets. The feature fusion method can acquire multi-scale features of images and capture objects of different sizes better. The combination of the two methods achieves satisfactory results.

5 Conclusions

We proposed a deep convolution neural network model in this paper, which utilizes deep supervision implicitly through the dense layer-wise connections and combines multi-level and multi-scale feature. Rich feature representation and feature reuse throughout the networks effectively improve the generalization ability of the detection model. The model has achieved state-of-the-art performance on two large-scale datasets for object detection in aerial images, under multiple settings. It may be a good feature extractor for various computer vision tasks and has certain reference value in engineering applications.

Acknowledgments The authors were supported in part by the National Natural Science Foundation of China under Grant 61702157, in part by NSF of Hebei Province through the Key Program under Grant F2016202144, in part by NSF of North China Institute of Aerospace Engineering through the Key Program under Grant ZD-2013-05, and in part by Self-financing Program of Langfang under Grant 2018013155.

References

1. D. Feng, A review on visualization of three-dimensional aerial corridor for low altitude safety. Electron. Meas. Technol. **41**(9), 2–9 (2018)
2. G.S. Xia, X. Bai, J. Ding et al., DOTA: a large-scale dataset for object detection in aerial images. arXiv, pp. 1–17
3. G. Cheng, P. Zhou, J. Han, RIFD-CNN: rotation-invariant and fisher discriminative convolutional neural networks for object detection. in *IEEE Conference on Computer Vision and Pattern Recognition(CVPR)*, 2016, pp. 2884–2893
4. Y. Long, Y. Gong, Z. Xiao, et al., Accurate object localization in remote sensing images based on convolutional neural networks. IEEE Trans. Geosci. Remote Sens. **55**(5), 2486–2498 (2017)
5. G. Wang, X. Wang, B. Fan, et al., Feature extraction by rotation-invariant matrix representation for object detection in aerial image. IEEE Geosci. Remote Sens. Lett. **14**(6), 851–855 (2017)
6. F. Zhang, B. Du, L. Zhang, et al., Weakly supervised learning based on coupled convolutional neural networks for aircraft detection. IEEE Trans. Geosci. Remote Sens. **54**(9), 5553–5563 (2016)
7. J. Deng, W. Dong, R. Socher et al., ImageNet: a large-scale hierarchical image database. in *IEEE Conference on Computer Vision and Pattern Recognition (CVPR)*, 2009, pp. 248–255
8. T. Lin, M. Maire, S. Belongie et al., Microsoft COCO: common objects in context. in *European Conference on Computer Vision (ECCV)*, 2014, pp. 740–755
9. G. Huang, Z. Liu, K.Q. Weinberger, Densely connected convolutional networks. in *IEEE Conference on Computer Vision and Pattern Recognition (CVPR)*, 2017, pp. 2261–2269
10. X. Yuan, D. Li, D. Mohapatra, M. Elhoseny, Automatic removal of complex shadows from indoor videos using transfer learning and dynamic thresholding. Comput. Electr. Eng. **70**, 813–825 (2018)
11. B.S. Murugan, M. Elhoseny, K. Shankar, J. Uthayakumar, Region-based scalable smart system for anomaly detection in pedestrian walkways. Comput. Electr. Eng. **75**, 146–160 (2019)
12. A. Krizhevsky, I. Sutskever, G.E. Hinton, ImageNet classification with deep convolutional neural networks. in *International Conference on Neural Information Processing Systems (NIPS)*, 2012, pp. 1097–1105
13. K. Simonyan, A. Zisserman, Very deep convolutional networks for large-scale image recognition. in *International Conference on Learning Representations (ICLR)*, 2015, pp. 1–13

14. C. Szegedy, W. Liu, Y. Jia et al., Going deeper with convolutions. in *IEEE Conference on Computer Vision and Pattern Recognition (CVPR)*, 2014, pp. 1–9

15. K. He, X. Zhang, S. Ren et al., Deep residual learning for image recognition. in *IEEE Conference on Computer Vision and Pattern Recognition (CVPR)*, 2016, pp. 770–778

16. R. Girshick. Fast R-CNN. in *IEEE International Conference on Computer Vision (ICCV)*, 2015, pp. 1440–1448

17. S. Ren, K. He, R. Girshick, et al., Faster R-CNN: towards real-time object detection with region proposal networks. IEEE Trans. Pattern Anal. Machine Intell. **39**(6), 1137–1149 (2017)

18. T.Y. Lin, P. Dollar, R. Girshick et al., Feature pyramid networks for object detection. in *IEEE Conference on Computer Vision and Pattern Recognition (CVPR)*, 2017, pp. 936–944

19. J. Redmon, A. Farhadi, YOLO9000: better, faster, stronger. in *IEEE Conference on Computer Vision and Pattern Recognition (CVPR)*, 2017, pp. 6517–6525

20. W. Liu, D. Anguelov, D. Erhan et al., SSD: single shot multibox detector. in *European Conference on Computer Vision (ECCV)*, 2016, pp. 21–37

21. K. Shankar, M. Elhoseny, R. Satheesh Kumar, S.K. Lakshmanaprabu, X. Yuan, Secret image sharing scheme with encrypted shadow images using optimal homomorphic encryption technique. J. Ambient. Intell. Humaniz. Comput. (2018). https://doi.org/10.1007/s12652-018-1161-0

22. K. Shankar, M. Elhoseny, S.K. Lakshmanaprabu, M. Ilayaraja, R.M. Vidhyavathi, M. Alkhambashi, Optimal feature level fusion based ANFIS classifier for brain MRI image classification. Concurrency Comput. Pract. Exp. 2018. https://doi.org/10.1002/cpe.4887

23. M. Elhoseny, G.-B. Bian, S.K. Lakshmanaprabu, K. Shankar, A.K. Singh, W. Wu, Effective features to classify ovarian cancer data in internet of medical things. Comput. Netw. **159**, 147–156 (2019)

24. N. Krishnaraj, M. Elhoseny, M. Thenmozhi, Mahmoud M. Selim, K. Shankar. Deep learning model for real-time image compression in Internet of Underwater Things (IoUT). J. Real-Time Image Process. 2019. https://doi.org/10.1007/s11554-019-00879-6

25. Y. Zhu, C. Zhang, D. Zhou, et al., Traffic sign detection and recognition using fully convolutional network guided proposals. Neurocomputing **214**, 758–766 (2016)

26. J. Long, E. Shelhamer, T. Darrell, *Fully convolutional networks for semantic segmentation.* in *IEEE Conference on Computer Vision and Pattern Recognition (CVPR)*, 2015, pp. 3431–3440

27. A. Takeki, T.T. Tu, R. Yoshihashi, et al., Combining deep features for object detection at various scales: finding small birds in landscape images. Trans. Comput. Vis. Appl. **8**(1), 5 (2016)

28. J. Wang, X. Wang, K. Zhang, et al., Small UAV target detection model based on deep neural network. J. Northwest. Polytech. Univ. **36**(2), 258–263 (2018)

29. Q. Lu, Y. Liu, J. Huang, X. Yuan, Q. Hu, License plate detection and recognition using hierarchical feature layers from CNN. Multimed. Tools Appl. **78**(11), 15665–15680 (2019)

30. X. Yuan, L. Xie, M. Abouelenien, A regularized ensemble framework of deep learning for cancer detection from multi-class, imbalanced training data. Pattern Recogn. **77**, 160–172 (2018)

31. X. Yuan, V. Sarma, Automatic urban water-body detection and segmentation from sparse ALSM data via spatially constrained model-driven clustering. IEEE Geosci. Remote Sens. Lett. **8**(1), 73–77 (2010)

32. G. Cheng, J. Han, P. Zhou, et al., Multi-class geospatial object detection and geographic image classification based on collection of part detectors. J. Photogramm. Remote Sens. **98**(1), 119–132 (2014)

33. H. Sun, X. Sun, H. Wang, et al., Automatic target detection in high-resolution remote sensing images using spatial sparse coding bag-of-words model. IEEE Geosci. Remote Sens. Lett. **9**(1), 109–113 (2011)

34. J. Han, P. Zhou, D. Zhang, et al., Efficient, simultaneous detection of multi-class geospatial targets based on visual saliency modeling and discriminative learning of sparse coding. J. Photogramm. Remote Sens. **89**(1), 37–48 (2014)

UAS Traffic Management in Low-Altitude Airspace Based on Three Dimensional Digital Aerial Corridor System

Dengchao Feng, Pengfei Du, Huiya Shen, and Zhenmin Liu

Abstract With the rapid development of the unmanned aerial system, the flight environment in low-altitude airspace is becoming more and more complex, which causes some potential security risks. A variety of air traffic management for low-altitude drones have emerged, among which three-dimensional aerial corridor system is a typical kind of low-altitude supervision mechanism with artificial intelligent theory. Firstly, the principle of three-dimensional aerial corridor system was illustrated. Then, the process of low-altitude traffic management based on aerial corridor system was described, the mainstream anti-drone technology and the visualization technology were discussed, and the relative regulation for air traffic management platform for low-altitude security industry was analyzed. Finally, the future work for air traffic management platforms based on aerial corridor systems was explored correspondingly, which is of positive significance for the further development of low-altitude security industry.

Keywords Unmanned aerial systems · Aerial corridor · Low-altitude airspace · Air traffic management

1 Introduction

With the opening of low-altitude airspace and the development of unmanned aerial systems (UAS) manufacturing industry, there are more and more types of UAS in low-altitude airspace [1], which increase the complexity of low-altitude flight environment and make it more difficult to operate and manage low-altitude airspace. Correspondingly, the low-altitude security issue has been widely concerned by all sectors of society [2]. In general, low-altitude security issues mainly focus on the following aspects, namely illegal theft of military intelligence, personal injury,

D. Feng (✉) · P. Du · H. Shen · Z. Liu
Department of Electronic and Control Engineering, North China Institute of Aerospace Engineering, Langfang, China

© Springer Nature Switzerland AG 2020
X. Yuan, M. Elhoseny (eds.), *Urban Intelligence and Applications*, Studies in Distributed Intelligence, https://doi.org/10.1007/978-3-030-45099-1_14

property damage, privacy infringement, transport of prohibited items and terrorist attacks, etc. [3].

In order to deal with the above problems, many countries have carried out a series of fruitful research and exploration work successively, among which a concept of UAS traffic management (UTM) system for low-altitude airspace seems more effective and has attracted the attention of some scientific research units, such as NASA and FAA in the USA, CAAC in China, etc. Statistic results show the UTM projects include 22 projects distributed in 11 countries in 2017 [4].

The future of UAS operations in low-altitude airspace depends in part on the efficient utilization of airspace resources and the safe operation and restriction of the flight plans. Accordingly, the structure of UTM platform can be generated based on a three-dimensional aerial corridor system in low-altitude airspace [5]. The construction of three-dimensional aerial corridor system was illustrated, among which the six key technologies were introduced briefly. The related international research status of air traffic management platform based on aerial corridor system was compared, the safety flight policies and regulations for UAS traffic platforms were discussed and the advanced anti-drone technologies in low-altitude security fields were explored correspondingly. The rest of this article is organized as follows: Section 2 introduces the principle of three-dimensional aerial corridor. Section 3 presents the structure of UTM platform based on three-dimensional aerial corridor system, the mainstream anti-drone technology, and the visualization technology. Section 4 analyzed the regulation of UTM platform. Section 5 concludes this paper with a summary and future work.

2 Principle of Three-Dimensional Aerial Corridor System

In order to ensure the healthy development of low-altitude security industry, UAS traffic management based on three-dimensional aerial corridor system has been advocated in recent years, which provides the low-altitude navigation information service, communication and information exchange service [6], air early warning and alarm services [7], illegal flight disposal service, etc. Three-dimensional aerial corridor system includes a variety of mainstream technologies, such as multivariate sensor detection technology, big data analysis technology, flight control technology, artificial intelligence based on deep learning [8–13], virtual reality, etc., which can provide a safe flight environment in low-altitude airspace, reduce the difficulty of low-altitude supervision, and ensure UAS safety in flight and take-off and landing.

Three-dimensional aerial corridor system involves six technical areas, namely delineation technology of aerial corridor in low-altitude airspace, spatial traffic network generation technology, space traffic routing planning technology, dynamic flight monitoring technology, UAS flight control technology, and key technologies of integrated service platform. Figure 1 shows the structure of three-dimensional aerial corridor system for UAS traffic management.

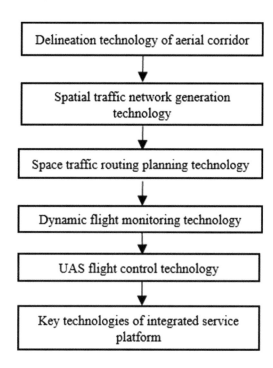

The target of delineation technology of aerial corridor is to ensure the seamless connection between surveillance airspace and reporting airspace in low-altitude airspace and realize the fast and maneuverable flight of UAS [14]. For spatial traffic network generation technology, the demand analysis and optimization procedure of air traffic network space can be executed based on spatial traffic network layout, topographic features and the regional distribution characteristics of residential, commercial, and industrial areas. The spatial traffic network search engine is created to realize the quantitative analysis and rapid construction of spatial traffic network. Space traffic routing planning technology focuses on the analysis of relevance between topography and spatial traffic network layout and realizes the static path planning and dynamic path planning [15]. Dynamic flight monitoring technology is used to promote the tracking and monitoring capabilities for various UAS and realize the real-time graphical recognition for flight message, flight path, and flight status. UAS flight control technology is used to design and simulate the flight control system, to modify flight path following strategy based on flight altitude, and ensures the independent, high precision, and high reliability to achieve task goals. The key technologies of integrated service include various communication interface, the related flight data management, space analysis, air traffic requirement forecasting model and visualized display, etc.

3   Construction of UAS Traffic Management Based on Aerial Corridor System

In order to deal with the potential security problems caused by UAS in low-altitude airspace, the UAS traffic management (UTM) based on the above three-dimensional aerial corridor system was proposed. The structure diagram of UTM based on the aerial corridor system is shown in Fig. 2.

Figure 2 shows the whole system of UTM consists of six layers, namely data acquisition layer, data service layer, an interface layer, platform layer, technical support layer, and the application layer. In data acquisition layer, a multi-sensor UAS detection network is formed by various detection devices, which include microwave detection equipment, passive radar detection equipment, voiceprint detector, optoelectronic tracking equipment, ADS-B devices, radio monitoring equipment, and miniature weather stations, etc. All the acquisition data can be transmitted to the data service layer. Data service layer includes data management, data mining, data storage, data backup, and data sharing. All the acquired data in data acquisition layer, such as radar data, ADS-B, GNSS, beacon data, image data, DSM and DTM are processed by a distributed parallel computing system in data service layer. The interface layer focuses on the interface schedule engine and hardware interface, among which the transmission protocol and gateway handoff are used to complete the switching function of public network and special network. In addition, the relative interface data generated in the interface layer is also transferred to the data service layer. The platform layer receives the data sets of

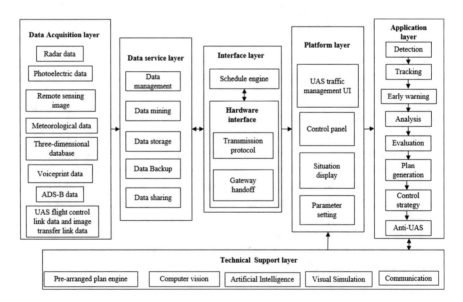

Fig. 2 The structure of UTM based on three-dimensional aerial corridor system

the interface layer and executed the data process. All the data sets are categorized by data type and sent to the relative function modules, which includes UI model of UAS traffic manage platform, the control panel, the UAS situation display, and various parameter setting model. The data process results of platform layer are sent to the application layer to realize the low-altitude environment detection and risk assessment, UAS traffic service and low-altitude defense for illegal UAS. The application layer mainly focuses on UAS detection, UAS tracking, early warning, flight environment analysis, and risk evaluation, flight plan generation and modification, UAS control strategy and anti-UAS technical application. The technical support layer mainly provides the relative theory and mathematical model for the platform layer and the application layer, which includes pre-arranged plan engine, computer vision theory, artificial intelligence, visual simulation, and communication technology.

According to the above structure of the UTM platform, there are three main operation process stages for UAS supervision, namely pre-flight preparation, real-time supervision during the flight and the low-altitude environment evaluation after the flight mission. In the pre-flight stage, the relative information needs to be provided, which includes UAS registration, flight plan submission, dynamic management of aerial corridor, aeronautical information services, metrological services, and flight information services [16]. In the flight stage, it mainly focuses on flight dynamic monitoring, flight interval service, emergency disposal, illegal flight investigation and punishment, and UAS traffic control support and overall coordination. After the flight mission, the whole flight status of UAS is analyzed and low-altitude environment is evaluated to ensure the security environment of flight area in low-altitude airspace.

For the illegal UAS in the flight area, the corresponding UAS defense technology is used in the UTM platform. Figure 3 shows the basic flowchart of anti-UAS technology in UTM platform for illegal UAS in low-altitude airspace. As can be seen from Fig. 3, it includes three stages, namely the UAS detection stage, UTM surveillance center, and UAS defense system stage. Various UAS detection data sets from UAS detection network are sent to the UTM surveillance center to execute the legitimacy identification of the drone. According to the hazard level, UTM platform starts the anti-UAS system.

For large-scale UAS, the relative ID data for each UAS is verified in the UTM platform, which includes the data of secondary radar transponder and the ID information of ADS-B device. If the verified data is not matched with the specified data of the platform, the UAS defense system is activated to protect the security environment of the specified area in low-altitude airspace. For the middle-scale UAS, the flight location information of UAS can be obtained by the public broadcasting mechanism based on ADS-B, which can be used to identify the legality of drone. For small-scale UAS, the data link monitoring technology is adopted to receive the data link location information to predict the potential threat areas for drone. For the illegal drone, the UAS defense system can be deployed in the specified area, such as electronic interference, sonic interference, electronic decoy, and electronic fencing, etc.

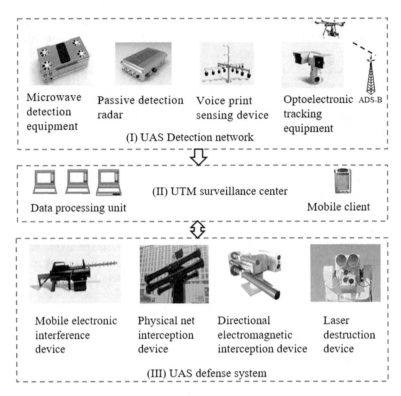

Fig. 3 Flowchart of anti-UAS in UTM platform

Fig. 4 Visualized scene of UAS flight area with an aerial corridor. (**a**) Aerial corridor in an urban area. (**b**) Aerial corridor in rural area

Finally, three-dimensional visualization technology is used in the whole UTM platform based on the aerial corridor system [17, 18], which includes three-dimensional visualization display of spatial real scene, spatial virtual scene, and multivariate heterogeneous parameter. Figure 4 shows some visualized scene in rural areas and urban areas.

The visualized UTM platform is an effective way to improve the efficiency of UAS supervision in low-altitude airspace, which allows the operators to track the connection between UTM operations and UAS management performance. By using big data visualization to monitor UAS, the airspace regulators can more easily detect the changes in the flight environment and adjust the regulatory solutions in time.

4 Analysis of UTM Regulation

The regulatory framework of UTM based on a three-dimensional aerial corridor includes two types of acceptable risk categories, namely the low-risk operation, high-risk operation. The operators need to evaluate the above operational risks, specify the operating restrictions, and provide a reasonable UAS operating plan to reduce the potential operational risk in low-altitude airspace.

Currently, one kind of categorization scheme for low-risk operation has been proposed by the ICAO to create a regulatory framework for UAS operations that starts with lower risk operations and moves to high-risk operations, which includes the low flying operations with the specified flight altitude range, a specific distance from the airport, building, and the crowds, the weather conditions suitable for flight and visual line of sight (VLOS) range, etc. Federal aviation administration (FAA) proposed the operational limitations for small UAS, such as the weight less than 55 lbs, daylight-only operations, "see-and-avoid" requirement, maximum airspeed of 100 mph, maximum altitude of 500 ft above ground level, minimum weather visibility of 3 miles from control station, and right-of-way to other aircraft, etc. European aviation safety authority (EASA) has released the regulatory framework for civil drones and an operations-centric, appropriate regulatory framework based on risk and performance. Civil aviation administration of China has released the administration document of operating flight activities of civilian unmanned aerial vehicles temporarily, which accelerates the progress of UTM regulation.

UTM regulation is of great value in ensuring the security of low-altitude in China. Currently, there are more than 200 UAS manufacturers and various UTM cloud platforms are put into trial operation. Some cities have also carried out pilot work of UTM to ensure low-altitude safe operation. In Shenzhen, UAS without the approval of the air force flight control department is prohibited to fly in the specified airspace.

Table 1 shows the main flight restriction in Shenzhen city, which includes three kinds of UAS, namely micro-drone, light-small UAS, and agricultural UAS for plant protection (Table 1).

According to the flight restriction of UAS in Shenzhen city, the operation of UTM platform based on aerial corridor systems needs to combine with the UAS regulation of local government. In order to ensure low-altitude security, some local cities also release UAS flight restriction documents temporally, such as Xi'an, Wuxi, Chongqing, etc. However, the flight environment of UAS in low-altitude airspace is complex, UAS operation often faces various subjective factors and objective factors caused by the uncertainty in different cities. Currently, there is some main

Table 1 Flight restriction of UAS in Shenzhen city

UAS type	Operation limitations
Micro-drone	Weigh less than 0.25 kg
	Maximum altitude of 50 m above ground level
	Maximum airspeed of 40 km/h
	No operations are allowed in no-fly zone and 2000 m range around the zone
	No operations are allowed in air danger zone and 1000 m around the zone
	No operations are allowed in the airport, temporary take-off and landing area, and 3000 m around the zone
	No operations are allowed in the range of 3000 m on the side of the national boundary line or actual control line
	No operations are allowed in the range of 100 m on the border side of Hong Kong to Shenzhen
	No operations are allowed in the prohibited military zone and the range of 500 m around it
	No operations are allowed in the government and supervisory office and the range of 200 m around the zone
	No operations are allowed in the satellite ground station and the range of 1000 m around the zone
	No operations are allowed in the meteorologic radar station and the range of 500 m around the zone
	No operations are allowed in the power plants, subway, gas station, wharf ports, large event area, and the range of 50 m around the zone
Light UAS	Self-weight less than 4 kg and the whole take-off weight less than 7 kg
	Maximum airspeed of 100 km/h
	Maximum altitude of 120 m above ground level
	No operations are allowed in no-fly zone and 5000 m range around the zone
	No operations are allowed in air danger zone and 2000 m around the zone
	No operations are allowed in the airport, temporary take-off and landing area, and 3000 m around the zone
	No operations are allowed in the range of 5000 m on the side of the national boundary line or actual control line
	No operations are allowed in the range of 500 m on the border side of Hong Kong to Shenzhen
	No operations are allowed in the prohibited military zone and the range of 500 m around it
	No operations are allowed in the government and supervisory office and the range of 500 m around the zone
	No operations are allowed in the satellite ground station and the range of 2000 m around the zone
	No operations are allowed in the meteorological radar station and the range of 1000 m around the zone
	No operations are allowed in the power plants, subway, gas station, wharf ports, large event area, and the range of 100 m around the zone
Agricultural plant protection UAS	Maximum altitude of 30 m above ground level
	Only allowed to operate above the government-mandated area of agroforestry and pastoral areas

concerned problem of UTM platform, which includes lack of standardization in UAS manufacturing and management, the different knowledge level of UAS operators, etc. Therefore, there is an urgent need to conduct a standardized study of UAS to ensure UAS meets the UTM requirement. In order to promote and standardize the UAS industry and meet the regulatory needs of the whole life cycle and ensure the security of low-altitude airspace, the government and some industry enterprise associations have carried out some research work in UAS standardization fields. For example, China national standardization management committee, the ministry of industry and technology, the department of science and industry, the civil aviation administration, and other eight ministries, have jointly issued the guidelines for the construction of the standard system for UAS in 2018, which has a positive effect on speeding up the construction of UTM platform based on aerial corridor system.

5 Conclusion

Low-altitude airspace is the closest area of airspace resources to the surface, which is a valuable strategic resource. It is an urgent task to guarantee the low-altitude security problem. UTM platform based on three-dimensional aerial corridor is used to realize the standardized supervision of UAS in low-altitude airspace. The principle of three-dimensional aerial corridor system was introduced and the structure of UTM platform based on three-dimensional was illustrated, the UTM regulation was analyzed, and the technical bottleneck issues were discussed correspondingly. UTM platform based on three-dimensional aerial corridor system adopts multi-disciplinary cross-cutting technology to realize the UAS supervision in low-altitude airspace, which is a complex giant system engineering that requires the participation of all relevant research institutions and management departments in society. It conforms to the urgent demand of the strategic development of the low-altitude security industry and has positive significance for promoting the healthy and sustainable development of low-altitude security industry. With the development of the advanced technology, such as artificial intelligence theory, industry big data technology, cloud computing, the UTM platform based on three-dimensional aerial corridor will show the trend of intelligent supervision for UAS and the network self-healing ability of UTM platform, and further improve the security supportability of UAS supervision in low-altitude airspace.

Acknowledgment This paper was supported by NCIAE postgraduate course teaching reform project with No.YJY201505 and the open scientific research fund of intelligent visual monitoring for hydropower project of three Gorges University with No.ZD2016106H.

References

1. D. Feng, X. Yuan, Advancement of security alarm chart visualization in low altitude airspace. J. Elect. Meas. Instrum. **29**(3), 305–316 (2015)
2. D. Feng, X. Yuan, Advancement of safety corridor and emergency management visualization in low altitude airspace. J. Electron. Meas. Instrum. **30**(4), 493–505 (2016)
3. D. Feng, H. Qin, The primary exploration of alarm chart visualization matching technology in low altitude airspace. Proc. IEEE ICEMI **1**, 1208–1213 (2015)
4. D. Feng, A review on visualization of three-dimensional aerial corridor for low altitude safety. Electron. Meas. Technol. **41**(9), 2–9 (2018)
5. D. Feng, Construction of aircraft traffic management platform for low altitude security based on three dimensional digital aerial corridor. Comput. Meas. Contr. **25**(12), 137–139 (2017)
6. D. Feng, A. Li, et al., A brief analysis of communication mode of android system in low altitude security monitoring data mobile display platform. Electron. Meas. Technol. **41**(9), 126–130 (2018)
7. D. Feng, H. Qin, Y. Zeng, Primary exploration of 3S technology in the matching design of visual alarm chart in low altitude airspace. Foreign Electron. Meas. Technol. **34**(6), 51–53 (2015)
8. K. Shankar, M. Elhoseny, R. Satheesh Kumar, S.K. Lakshmanaprabu, X. Yuan, Secret image sharing scheme with encrypted shadow images using optimal homomorphic encryption technique. J. Ambient. Intell. Humaniz. Comput. 2018. https://doi.org/10.1007/s12652-018-1161-0
9. K. Shankar, M. Elhoseny, S.K. Lakshmanaprabu, M. Ilayaraja, R.M. Vidhyavathi, M. Alkhambashi. Optimal feature level fusion based ANFIS classifier for brain MRI image classification. Concurrency Comput. Pract. Exp. 2018. https://doi.org/10.1002/cpe.4887
10. M. Elhoseny, G.-B. Bian, S.K. Lakshmanaprabu, K. Shankar, A.K. Singh, W. Wu, Effective features to classify ovarian cancer data in internet of medical things. Comput. Netw. **159**, 147–156 (2019)
11. N. Krishnaraj, M. Elhoseny, M. Thenmozhi, M.M. Selim, K. Shankar. Deep learning model for real-time image compression in Internet of Underwater Things (IoUT). J. Real-Time Image Process. 2019. https://doi.org/10.1007/s11554-019-00879-6
12. B. Fang, X. Guo, Z. Wang, Y. Li, M. Elhoseny, X. Yuan, Collaborative task assignment of interconnected, affective robots towards autonomous healthcare assistant. Futur. Gener. Comput. Syst. **92**, 241–251 (2019)
13. X. Yuan, D. Li, D. Mohapatra, M. Elhoseny, Automatic removal of complex shadows from indoor videos using transfer learning and dynamic thresholding. Comput. Electr. Eng. **70**, 813–825 (2018)
14. D. Feng, Construction of aerial corridor for unmanned aircraft systems in low altitude airspace based on point cloud of laser scanner. Comput. Meas. Control **26**(2), 133–140 (2018)
15. D. Feng, X. Yuan, L. Kong, Research on the visual flight planning of UAS for privacy protection in low altitude airspace. Proc. Int. Symp. Test Autom. Instrum. **1**, 240–244 (2016)
16. D. Feng, L. Liang, et al., Research on the structure of low altitude monitoring system for unmanned aerial system. Electron. Meas. Technol. **41**(9), 141–145 (2018)
17. D. Feng, X. Yuan, Digital terrain model extraction in SUAS clearance survey using Lidar data. Proc. IEEE IGARSS **1**, 791–794 (2016)
18. D. Feng, X. Yuan, Automatic construction of aerial corridor for navigation of unmanned aircraft systems in class G airspace using Lidar. Proc. SPIE Def. Commer. Sens. **1**, 47–49 (2016)
19. Civil Aviation Administration of China. Implementation of UAV flight management in Shenzhen. http://www.caac.gov.cn/local/ZNGLJ/ZN_XXGK/ZN_ZWGG/201811/t20181116_192933.html?from=singlemessage

Analyzing of Spatial Interactive Network Based on Urban Community Division

Ning Li, Yaqin Ye, Jiao Pan, Yingqiang Zhong, and Qiao Hua

Abstract The urban community interaction patterns are an important portrait of the urban spatial structure, and they can serve with urban community construction, traffic management, and resource allocation. In the big data era, various movement trajectories are available for studying spatial structures. In this study, on the basis of the massive taxi-trip data of Wuhan, we built a spatially embedded network and identified the intra-city spatial interactions. The community detection method was applied to reveal urban structures of different times in Wuhan. At the same time, we studied the degree of association between different regions based on the frequency of interactions of trajectory data. Finally, we found that: (a) Compared to weekends and working days, people have a wider range of travel and more random travel locations on holidays. (b) From community detection, Hanyang District and Hankou Districts were classified as the same "community," and the result of Wuchang District division was similar to the administrative boundaries. (c) In Wuhan, the most closely related areas were Hankou and Wuchang District, and the closeness between Wuchang District and Qingshan District was the second. From the research results, it can be concluded that the closeness of community interaction is positively related to the level of regional economic development, which illustrates the importance of community development and provides decision-making basis for urban traffic management.

Keywords Urban structure · Community detection · Spatially embedded network · GPS taxi data · Travel pattern

1 Introduction

Urban spatial structures include urban morphology and urban interaction. The urban morphology refers to the spatial distribution pattern of various elements of the

N. Li · Y. Ye (✉) · J. Pan · Y. Zhong · Q. Hua
Faculty of Information Engineering, China University of Geosciences, Wuhan, China

© Springer Nature Switzerland AG 2020
X. Yuan, M. Elhoseny (eds.), *Urban Intelligence and Applications*, Studies in Distributed Intelligence, https://doi.org/10.1007/978-3-030-45099-1_15

city (including physical facilities, social groups, economic activities, and public institutions). The urban interaction refers to the interrelationship between the elements of the city, which integrates the behavior of group activities into entities with different functions [1]. A reasonable urban spatial structure can shorten the flow space and time of people flow, object flow, information flow, and capital flow. It can also alleviate urban pollution, traffic congestion, urban heat island effect, and other issues. Besides, it can improve urban resource utilization and economic benefits [2]. The characteristics of community gathering based on residents' travel and activities can directly reflect the urban socio-economic function layout. Exploring the community aggregation patterns and spatial interaction characteristics of the city in different periods and combining these different spatial characteristics are of great significance for understanding the activities of urban residents and the spatial structure of urban.

With the advent of the era of big data, a large number of emerging geo-tag data was processed, such as mobile phone communication data, taxi GPS data, intelligent transportation card data, social media data, and so on [3–7]. Because people's movement in urban has predictable and mobile [8], using the spatial interactivity of these data to explore the natural boundaries and regional structures formed by the social behavior of residents in a city has become a new research hotspot. Specifically, the spatial interactions refer to the actual movements of people in the urban. For example, many researchers have used these trajectory data, which are the spatial footprints of citizens' activities, revealing urban structure and trying to determine whether existing administrative boundaries are still reasonable [9–12], comparing the evolutionary characteristics of urban structure in time and space [13], summarizing the travel patterns of urban residents [14], and providing suggestions for regional partitions [15].

In these studies, researchers introduced the methods of network science into the analysis of geographic information. The spatial region is generally considered to be a node in the network. The spatial interaction between the residents in the city interacts to represent the edges in the network, and the frequency of interaction was considered to be an edge with different weights. These network nodes with spatial location information can reflect the impact of geospatial factors on staff mobility. By dividing the network space, based on the comparative analysis of urban community structure and administrative division, the spatial area is divided into different communities. It is possible to analyze the relationship between different urban areas and the collective flow patterns between different regions. At the same time, it showed that large-scale trajectories and behavioral data of resident activity are providing new ways to explore the characteristics of urban areas and the aggregation characteristics of spatial interactions. Most of these studies pay more attention to the similarities and differences between the results of regional division and traditional administrative boundaries, and pay little attention to the degree of association between regions, and also lack the division of time for fineness. The degree of relevance is closely related to traffic travel and urban planning. Studying the travel data of urban residents at different time periods can also explain the urban spatial structure from different angles and explore the law of residents' travel, which is of great significance.

The taxi GPS data have been widely used to investigate urban structure since the 1970s [16] to implement our research. A large number of taxi GPS trajectory data greatly promoted the development of related research, including accessibility analysis of urban road network [17, 18], urban planning [19, 20], the research on hot spots [21–24],transportation analysis [25–27], land use classification [28–30], and human mobility patterns [31, 32]. As important public transport, taxies have the characteristics of long operating time, wide coverage, and a large degree of freedom, and its trajectory data contains characteristics such as urban functional structure, urban traffic evolution model, and residents' travel activities. With the information of when and where passengers are picked up or dropped off by taxi, the spatio-temporal events of residents' travel can be profoundly described and meaningful trips corresponding to displacements between people's consecutive activities are easy to extract, which requires a difficult and time-consuming effort to obtain from other forms of data. Precise spatio-temporal properties of massive intra-city trips generated by taxi trajectories lay a solid foundation to reflect completely the structure of the city.

In our research, we introduced complex network science methods to analyze taxi trajectory data of Wuhan, and explored the urban spatial structure at different times. Firstly, a uniform grid was set as the basic geographic unit in the entire area of Wuhan, in which each unit acts as a node within the network, taking the taxi trajectory. The data was moved to establish a connection between the grid cells to form a connected network. We used the map equation algorithm [33] to divide the network and visualize the results of the partitioning. Secondly, we used the taxi trajectory frequency to simulate the degree of association between different regions. Finally, we extracted the peak period of urban residents' travel, and analyzed the aggregation characteristics at different times of the city, and summarized the travel rules. It also provided new insights into the spatial interaction between human and urban structures.

2   Research Areas and Data

2.1   Research Area

As the capital of Hubei Province, Wuhan is located in the middle and lower reaches of the Yangtze River. It is the most populous city in Central China. The total area of Wuhan is about 8494.41 km^2, and the total population in 2014 was 10.2 million. The city has jurisdiction over 13 districts and counties, including 7 core areas, namely Jiang'an, Jianghan, Qikou, Hanyang, Qingshan, Wuchang, and Hongshan (Fig. 1). In the past few decades, Wuhan has experienced a period of rapid urbanization. According to the 2013 annual report of Wuhan Municipal

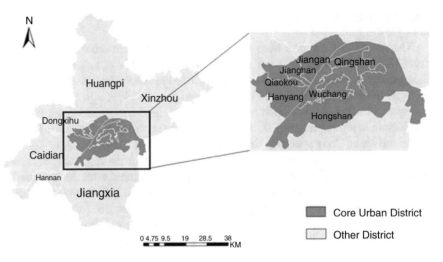

Fig. 1 Map of the Wuhan city that is the study area of this work

Ministry of Communications, the traffic volume in Wuhan reached 2.07 billion in 2012. In addition, the passenger volumes of the bus, rail transit, taxi, and ferry account for 76.7%, 4.1%, 18.7%, and 0.5% of total passenger volume, respectively. Therefore, taxi travel is an important component of public transport in Wuhan, and taxi trajectories provide a reasonable source of data for urban research because they have the ability to capture most of the city's passenger flow.

2.2 Taxi Trajectory Data and Data Preparation

The GPS trajectory data set in this paper contains GPS trajectory data generated by 12,529 taxis in Wuhan from May 1 to May 10, 2014. The data collection interval is about 1 s, and the daily data set contains a total of nearly 15 million records. The data mainly records 10 fields of taxi ID, latitude, longitude, time, vehicle speed, driving direction, and passenger status.

Because on different dates, urban residents will have different travel patterns, which will lead to different aggregation characteristics of urban community structure. Holidays, workdays, and weekends are three basic date types in a year. Therefore, we selected data from May 1, 2014 (holiday), May 5 (working day), and May 10 (weekend) for 3 days as an example, and explored the different urban structures caused by the spatial interaction of travel trajectories during holidays, working days, and weekends.

We deleted data whose latitude and longitude are not within the scope of Wuhan and checked the latitude and longitude range of the original data before using the data for operation. The latitude and longitude of Wuhan City is longitude 113.41–

115.05, the latitude is 29.58–31.27. And, we deleted the duplicate operation. Due to some equipment and network failures in data acquisition, there are several track points at the same time and the same position on the same track. In order to perform the test well, we found out the duplicate data to delete.

2.3   Map Matching

The road network in this article is from the official website of OpenStreetMap (OSM), a website that provides free access to road networks. The coordinate system of the road network is WGS84, and the taxi trajectory data is the Mars coordinate system. Therefore, the coordinate system of the taxi trajectory data is converted to WGS84 before the map matching is performed. The map matching will match the latitude and longitude of the GPS trajectory data with the road network, to find the real position of the taxi trajectory on the road section at the moment, and to improve the position accuracy of the taxi data. The map matching based on proximity analysis is specifically described in Fig. 2. Points A1, A2, A3..., A11 form a trajectory, and points A4, A5, and A9 are not mapped to the corresponding road. Map matching is to move a point that does not match the electronic map to the nearest road or delete it as a noise point. Among them, two points A4 and A5 are moved to the nearest road by corresponding operations, and A9 points are deleted as noise points. The easiest way to solve map matching is to locate the GPS track point on the nearest road. We used the proximity analysis tool in ArcMap to operate for this study.

Importing the two types of data into ArcGIS reveals that the track points are inconsistent with the road matching, resulting in errors in the GPS track with some

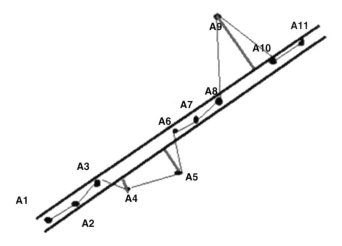

Fig. 2 Map matching based on proximity analysis

locations on the actual road. In order not to affect the subsequent experiments, we matched the trajectory data with the electronic map.

3   Method

3.1   Building Spatial Interaction Networks

In the complex network model, we can represent the network with $G = <V, E, W>$, where V is a set of spatial nodes corresponding to the underlying urban regions, E is a set of edges each representing the connection between a pair of nodes, and the W is assigned by the accumulated volume of people's movements. The entire network is constructed as a directed weighted network.

In real life, the movement of taxis is defined as the geographical migration or movement of individuals [34], where people move in cities to connect discrete locations into a comprehensive system. In order to identify clusters of connectivity in urban areas, an interactive connection network must be formed. In this study, in order to establish a network using taxi trajectory data, firstly, we divided the entire research area into uniform basic units and used the grid to represent the node of the network. Secondly, we extracted the origin point and decision point (OD) of the taxi trajectory data. If a taxi goes from one node to another, a directed edge was set between the two nodes. Last, the spatial interaction between each grid was counted, and the number of tracks between nodes is used as the number of tracks. Based on this, a grid-based directed weighted network that can be represented by the G mentioned above was established. The whole process is shown in Fig. 3. In our study, the basic unit size of the grid is 1×1 km. The size is selected according to the relevant research [35], the size of 1×1 km can describe the urban structure in detail.

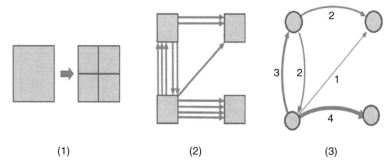

(1)   (2)   (3)

Fig. 3 Building a spatial interactive network based on grids

3.2   Community Structure of the Network of Spatial Interactions

An interactive network was constructed through taxi trajectory data, which is a directed weighted graph. We further determined clusters of strongly connected spatial nodes, which are known as communities in the context of graph space. There are a variety of community detection algorithms that produce different results. Infomap algorithm was used because it is suitable for applying to directed weighted graphs [36] and it has good efficient.

Infomap algorithm identifies communities by minimizing the expected length of the trajectory of a random walker, the idea of this algorithm is looking for a modular map M that partitions the n nodes into m modules so as to minimize the expected description length of a random walk. Using map M, the average description length of a single step is given by [37]:

$$L(M) = q H(Q) + \sum_{i=1}^{m} p_i H (p_i) \tag{1}$$

where $qH(Q)$ is the entropy of movement among clusters and $\sum_{i=1}^{m} p_i H (p_i)$ is the entropy of movement within clusters. Specifically, q is the probability that a random walker jumps from one cluster to another, while p_i is the probability of movement within cluster i. i and m are integers, and i is smaller than or equal to m. The goal of the network community is to find the partition that minimizes the average description length $L(M)$ as the optimal community partition in all possible community divisions. The specific process is to use the greedy algorithm to find the optimal partition, that is, to initially assign an independent community to each node, and merge the two communities with the average description length $L(M)$ to decrease the most, repeat the process until the final merge into a community.

4   Results

4.1   Collective Mobility Patterns of People in Wuhan

As mentioned above, we had divided the entire city of Wuhan into a uniform grid of 1 × 1 km on a spatial scale. On the time scale, the 3 days of May 1st, May 5th, and May 10th were selected as the research objects, representing the three periods of holidays, working days, and weekends. Before revealing the spatial structure of the city, we studied the collective flow patterns of people at different times. People will have different modes of travel in three time periods. For example, we usually have more regular daily activities during workdays and periods, and commuting between family and workplace accounts for a large part of daily travel; during weekends, people may be more concentrated in entertainment; during the holidays, people may go out for traveling, thus resulting in more random trajectories. In order to explore the statistical characteristics of people traveling in three different periods, we introduced the degree distribution and network diameter of a complex network. In a complex network, the degree distribution refers to the number of times connected to the node, and the network diameter represents the length of the edge. The indicators were used to simulate the movement frequency and the movement radius. We calculated the complementary cumulative distribution function (CCDF) of the movement frequency and the movement radius for three different periods, as shown in the following Fig. 4.

As can be seen from Fig. 5, three different time periods are compared. In the movement frequency, the working day is equal to the weekend and is greater than the holiday. In the movement radius, the holiday is greater than the working day and is equal to the weekend. The reason for our analysis may be that the travel locations on working days and weekends are more fixed. During the holidays, as the number of foreign tourists in the city increases, the trajectory they generate is more random, and the citizens are more inclined to travel long distances. This is also the same as our traditional perception.

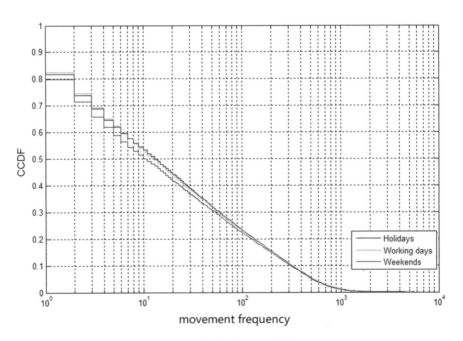

Fig. 4 Cumulative distribution function (CDF) of movement frequency

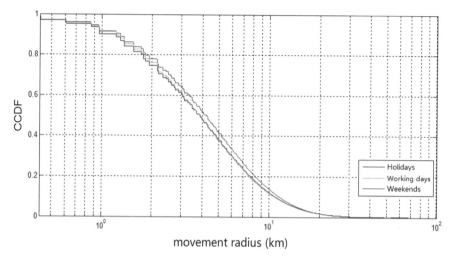

Fig. 5 Cumulative distribution function (CDF) of movement radius

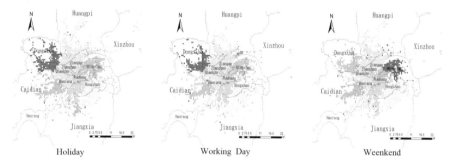

Fig. 6 The size of the cell is 1×1 km

4.2 Revealing the Urban Structure of Wuhan

According to the method mentioned above, we used the data from the entire Wuhan city on holidays, workdays, and weekends to build a spatial interaction network. The network node is about 2000, and the number of sides is about 260,000. The results of the regional division are shown in Fig. 6.

As shown in Fig. 6, it can be seen that the results of the regional division show that the entire Wuhan city is roughly divided into 6–7 communities, and most of the geographical divisions are concentrated in the central districts with relatively dense population, concentrated in Wuhan. Some remote districts such as Xinzhou District, Hannan District, Huangpi District have a large blank space. Thus, the core area of Wuhan can be verified. At the same time, it also shows that Wuhan is a typical multi-center city separated by the Yangtze River. The boundary of the interaction between the residents' activities and the administrative division from the taxi data overlaps in some places, but there are some differences in other places. In the results, it is shown that in the three periods, Hanyang and Hankou Districts both are classified as the same "community," then it can be inferred that the population flow between the two places is more frequent. The result of Wuchang division is similar to the administrative boundaries, it may be related to a large proportion of college students in Wuchang's demographic composition. On the time scale, during the holidays, compared to working days and weekends, more subtle areas can be detected, such as the Huanghua District, Xinzhou District, and the clusters in the small center can be seen very intuitively in the picture; during the weekends, a new "community" was detected in the middle of Wuchang District and Qingshan District. From the comparison of remote sensing images, there are many universities and commercial land in this area. We analyzed that during the weekend, people may prefer to travel inside.

4.3   Measure the Degree of Association Between Regions

For the results of the region division, we want to further simulate the interaction information between regions and measure the degree of association between modules. In our study, we proposed a measure of the correlation coefficient between regions Q:

$$Qc, v = Mc, v/M \tag{2}$$

In Eq. (2), C and V represent different regions, respectively, and M represents the number of OD matrices between regions. If there are more interactions between the two regions, it means that the traffic between the two regions is closer, and the correlation coefficient Q between the two modules will also be bigger. When $C = V$, Q represents the tightness inside the module. For holidays and weekdays, we measure the degree of association between the top six regions. For the weekend, we measure the correlation coefficient between the top seven regions, as shown in Fig. 7.

The final result is shown in Fig. 8. By comparing the module coefficients between different communities over three time periods, we can see that on non-working days, the correlation between communities 12 and 14 is significantly greater than the working day. It can be concluded that on non-working days, people travel relatively more across the river, people tend to travel longer distances, and on weekdays people may tend to travel within the region. At the same time, according to the correlation coefficient between modules, we can see that the three modules with the largest correlation coefficient are 12, 23, 24, and the analysis can be obtained. In Wuhan, the most closely related is Hankou-Hanchang District and Wuchang District. The closeness between Wuchang District and Qingshan District was the second.

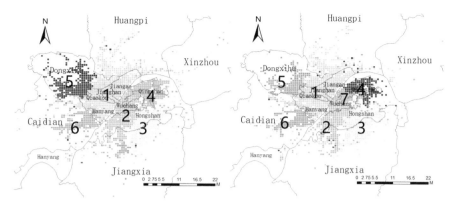

Fig. 7 Number of each region

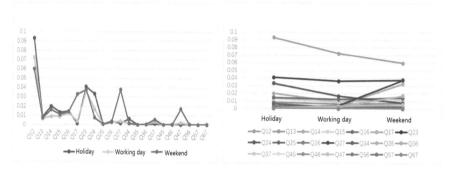

Fig. 8 The correlation coefficient between different regions

4.4 Urban Structure at Different Time Periods

To further study the internal structure of the city in different time periods. First, we selected the peak period of urban residents' travel, because the passengers' pick-up points and drop-off points in the trajectory data reflect the residents' travel status. We statistically analyzed the number of passengers' pick-up and drop-off points to get the time-space distribution of the getting-off data. We randomly selected holidays, workdays, and weekends for 100 taxis per day for three periods, counting the number of trains and the number of trips, and the results are shown in Fig. 9.

There are obviously several peak trips in 3 days. We chose a total of 8:00–10:00, 13:00–14:00, 18:00–19:00, and 23:00–24:00 to analyze during peak hours. The data in the corresponding time period was selected to construct an interactive network, and community detection was performed for three different time periods to infer the urban structure (Figs. 10 and 11).

According to the results of community testing at the peak of four travel periods in three different time periods, the urban structure revealed by travel data in a period of

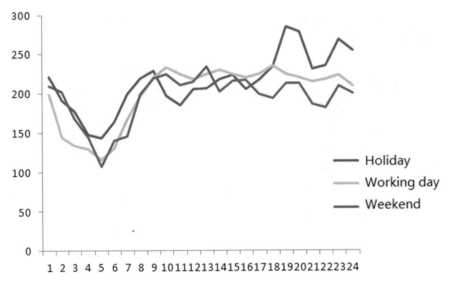

Fig. 9 Change in the number of passengers on different dates

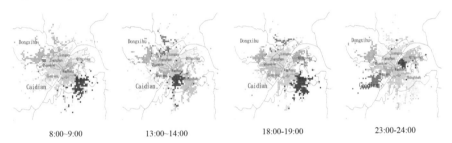

8:00~9:00   13:00~14:00   18:00-19:00   23:00-24:00

Fig. 10 Aggregation characteristics in four periods during the holiday

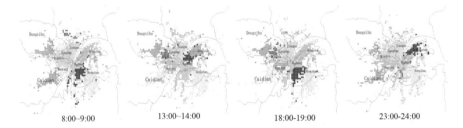

8:00~9:00   13:00~14:00   18:00-19:00   23:00-24:00

Fig. 11 Aggregation characteristics in four periods during the working day

time is more elaborate than the urban structure revealed by the travel data in a whole day. We found that, during the period from 8:00 to 9:00, the community structure and administrative boundaries in the working day are more similar, for example, Wuchang District and Hongshan District are divided into different "communities."

8:00–9:00   13:00–14:00   18:00–19:00   23:00–24:00

Fig. 12 Aggregation characteristics in four periods during the weekend

Because during the working day, this period is the peak period of work, people's travel patterns tend to be "residential areas—office areas." During non-working days, people's travel patterns are more random, resulting the urban structure is very different from the administrative boundary; during the period from 13:00 to 14:00, the Qiaokou District is detected as an independent "community" in three periods; during the period from 18:00 to 19:00, the number of "communities" was more, and the gathering of "communities" was more dispersed; during the period from 23:00 to 24:00, the "community" gathering is more regular, and the category of "community" is also less, because, during this period, the travel patterns of urban residents are usually "recreational areas—residence areas," the number of random tracks generated is less. From the perspective of the whole result, Jianghan District and Jiang'an District are classified as the same "community," and the two regions have the greatest degree of association and the greatest degree of association. From the four periods of peak travel time, during the period from 18:00 to 19:00, the regional gathering is most similar to the administrative boundary. During the period from 23:00 to 24:00, the regional gathering is the most regular (Fig. 12).

5   Conclusion and Future Work

In this study, we used the GPS trajectory data of taxi as the research object from the perspective of time and space to build the spatial interactive network. We introduced the network science method, especially the community detection method, to reveal regional urban structure on the basis of spatial interaction. We compared the results with the administrative boundaries delineated by traditional urban planners to the analysis of similarities and differences. At the same time, on the basis of regional division, we measured the correlation coefficient between different communities to find out more relevant communities. Finally, we inferred the urban structure in a more detailed way according to the peak period of travel of urban residents and compared it with administrative boundaries to better understand the changes in the urban structure.

The urbanization process in China has been rapid, the urban structure of second-tier cities such as Wuhan is becoming more complex. Big data provides us with the opportunity to understand the relationship between resident mobility patterns and corresponding urban structures, laying the foundation for new theoretical research for the city and transportation planning has contributed. The methods provided in this study are also applicable to analysis in other cities. Further research may expand data sources, including private cars, buses, subway trips, social media travel data, etc. This combination of diverse data can more describe human mobility and urban structure, providing a multifaceted view of urban dynamics. We can also extend the data source in the time dimension. The availability of long-term mobility data would make it possible to detect changes in the urban structure and verify the effectiveness of the policy.

References

1. L.S. Bourne, *Internal Structure of the City: Reading on Urban Form, Growth and Policy* (Oxford University Press, Oxford, 1982)
2. D. Lu et al., *Regional Development and Its Spatial Structure* (Science and Technology Press, Beijing, 1998)
3. H. Rao, X. Shi, A.K. Rodrigue, J. Feng, Y. Xia, M. Elhoseny, X. Yuan, L. Gu, Feature selection based on artificial bee colony and gradient boosting decision tree. Appl. Soft Comput. (2018). https://doi.org/10.1016/j.asoc.2018.10.036
4. K. Shankar, M. Elhoseny, R. Satheesh Kumar, S.K. Lakshmanaprabu, X. Yuan, Secret image sharing scheme with encrypted shadow images using optimal homomorphic encryption technique. J. Ambient. Intell. Humaniz. Comput. (2018). https://doi.org/10.1007/s12652-018-1161-0
5. M. Elhoseny, X. Yuan, Z. Yu, C. Mao, H. El-Minir, A. Riad, Balancing energy consumption in heterogeneous wireless sensor networks using genetic algorithm. IEEE Commun. Lett. **19**(12), 2194–2197 (2015). https://doi.org/10.1109/LCOMM.2014.2381226
6. N. Krishnaraj, M. Elhoseny, M. Thenmozhi, Mahmoud M. Selim, K. Shankar, Deep learning model for real-time image compression in Internet of Underwater Things (IoUT). J. Real-Time Image Process. 2019. https://doi.org/10.1007/s11554-019-00879-6
7. B.S. Murugan, M. Elhoseny, K. Shankar, J. Uthayakumar, Region-based scalable smart system for anomaly detection in pedestrian walkways. Comput. Electr. Eng. **75**, 146–160 (2019)
8. C. Song et al., Limits of predictability in human mobility. Science **327**(5968), 1018–1021 (2010)
9. C. Ratti, S. Sobolevsky, F. Calabrese, et al., Redrawing the map of Great Britain from a network of human interactions. PLoS ONE **5**, 0014248 (2010)
10. S. Rinzivillo, Discovering the geographical borders of human mobility. Künstliche Intell. **26**(3), 253–260 (2012)
11. S. Gao, Y. Wang, Y. Gao, Y. Liu, Understanding urban traffic-flow characteristics: a rethinking of betweenness centrality. Environ. Plan. B Plan. Des. **40**, 135–153 (2013)
12. J. Yin, A. Soliman, D. Yin, et al., Depicting urban boundaries from a mobility network of spatial interactions: a case study of Great Britain with geo-located Twitter data. Int. J. Geogr. Inf. Sci. **31**(7), 1293–1313 (2017)
13. C. Zhong, X. Huang, M. Batty, et al., Detecting the dynamics of urban structure through spatial network analysis. Int. J. Geogr. Inf. Sci. **28**(11), 2178–2199 (2014)
14. Y. Tanahashi, J.R. Rowland, S. North, et al., Inferring human mobility patterns from anonymized mobile communication usage, in *International Conference on Advances in Mobile Computing & Multimedia*, vol. 2012 (ACM, New York, 2012), pp. 151–160

15. A.D. Montis, S. Caschili, A. Chessa, Commuter networks and community detection: a method for planning sub regional areas. Eur. Phys. J. Spec. Top. **215**(1), 75–91 (2013)

16. J.B. Goddard, Functional regions within the city centre: a study by factor analysis of taxi flows in central London. Trans. Inst. Br. Geogr. **49**, 161–182 (1970)

17. Q. Li, Z. Zeng, T. Zhang, et al., Path-finding through flexible hierarchical road networks: an experiential approach using taxi trajectory data. Int. J. Appl. Earth Obs. Geoinf. **13**(1), 110–119 (2011)

18. J. Cui, F. Liu, J. Hu, D. Janssens, G. Wets, M. Cools, Neurocomputing identifying mismatch between urban travel demand and transport network services using GPS data: a case study in the fast growing Chinese City of Harbin. Neurocomputing **181**(12), 4–18 (2016)

19. Y. Zheng, Y. Liu, J. Yuan, X. Xie, Urban computing with taxicabs, in *Proceedings of the 13th International Conference on Ubiquitous Computing* (ACM, New York, 2011), pp. 89–98

20. M. Veloso, S. Phithakkitnukoon, C. Bento, Sensing urban mobility with taxi flow, in *Proceedings of the 3rd ACM SIGSPATIAL International Workshop on Location-Based Social Networks* (ACM, New York, 2011), pp. 41–44

21. Y. Yue, Y. Zhuang, Q. Li, Q. Mao, Mining time-dependent attractive areas and movement patterns from taxi trajectory data, in *17th International Conference on Geoinformatics, 2009* (IEEE, Fairfax, 2009), pp. 1–6

22. H.W. Chang, Y.C. Tai, J.Y.J. Hsu, Context-aware taxi demand hotspots prediction. Int. J. Bus. Intel Data Min. **5**(1), 3–18 (2009)

23. B. Li, D. Zhang, L. Sun, C. Chen, S. Li, G. Qi, Q. Yang, Hunting or waiting? Discovering passenger-finding strategies from a large-scale real-world taxi dataset, in *IEEE International Conference on Pervasive Computing and Communications Workshops (PERCOM Workshops)* (IEEE, Seattle, 2011), pp. 63–68

24. P. Zhao, K. Qin, X. Ye, Y. Wang, Y. Chen, A trajectory clustering approach based on decision graph and data field for detecting hotspots. Int. J. Geogr. Inf. Sci. **31**(6), 1101–1127 (2017)

25. Z. Fang, S.-L. Shaw, W. Tu, Q. Li, Y. Li, Spatiotemporal analysis of critical transportation links based on time geographic concepts: a case study of critical bridges in Wuhan. Chin. J. Transp. Geogr. **23**, 44–59 (2012)

26. S. Gao, Y. Wang, Y. Gao, Y. Liu, Understanding urban traffic-flow characteristics: a rethinking of betweenness centrality. Environ. Plan. B Plan. Des. **40**, 135–153 (2013)

27. Q. Li, T. Zhang, H. Wang, Z. Zeng, Dynamic accessibility mapping using floating car data: a network-constrained density estimation approach. J. Transp. Geogr. **19**, 379–393 (2011)

28. C. Kang, S. Sobolevsky, C. Ratti, Y. Liu, Exploring human movements in Singapore: a comparative analysis based on mobile phone and taxicab usages. in *Urb Comp'13. Chicago, Illinois, USA*, 2013

29. G. Pan, G. Qi, Z. Wu, D. Zhang, S. Li, Land-use classification using taxi GPS traces. IEEE Trans. Intell. Transp. Syst. **14**(1), 113–123 (2013)

30. X. Liu, C. Kang, L. Gong, et al., Incorporating spatial interaction patterns in classifying and understanding urban land use. Int. J. Geogr. Inf. Sci. **30**(2), 1–17 (2015)

31. X. Liang, X. Zheng, W. Lv, T. Zhu, X. Ke, The scaling of human mobility by taxis is exponential. Phys. A Stat. Mech. Appl. **391**(5), 2135–2144 (2012)

32. Y. Xu, S.-l. Shaw, Z. Zhao, L. Yin, F. Lu, J. Chen, X. Yang, et al., Another tale of two cities:understanding human activity space using actively tracked cellphone location data. Ann. Assoc. Am. Geogr. **106**(2), 489–502 (2016)

33. M. De Domenico et al., Identifying modular flows on multilayer networks reveals highly overlapping organization in interconnected systems. Phys. Rev. X **5**(1), 011027 (2015)

34. M.C. González, C.A. Hidalgo, A.-L. Barabási, Understanding individual human mobility patterns. Nature **453**, 779–782 (2008)

35. Liu Y, Wang F, Xiao Y, et al. Urban land uses and traffic 'source-sink areas': evidence from GPS-enabled taxi data in Shanghai. Landsc. Urban Plan., 2012, 106(1):0-87.

36. A. Lancichinetti, S. Fortunato, J. Kertész, Detecting the overlapping and hierarchical community structure in complex networks. N. J. Phys. **11**(3), 033015 (2009)

37. M. Rosvall, D. Axelsson, C.T. Bergstrom, The map equation. Eur. Phys. J. Spec. Top. **178**(1), 13–23 (2010)

Part IV
Security, Safety, and Emergency Management

Measuring Vulnerability for City Dwellers Exposed to Flood Hazard: A Case Study of Dhaka City, Bangladesh

Md. Enamul Huq, A. Z. M. Shoeb, Akib Javed, Zhenfeng Shao, Mallik Akram Hossain, and Most. Sinthia Sarven

Abstract Recently, floods and natural disasters are occurring over the world frequently. It is not possible to avoid the natural occurrence of floods, but it could be managed effectively or minimize the losses and damages with the appropriate technique. The purpose of this study is to measure household vulnerability to flood hazard quantitatively as a tool for mitigation aspects. The study was conducted based on primary data. Therefore, 300 households (150 from slum and 150 from non-slum) were surveyed with a structured questionnaire on responsible factors (social, economic, institutional, structural, and environmental) of household vulnerability to flooding. The relative weight of each variable and indicators were assigned with the Analytical Hierarchy Process (AHP) to obtain household vulnerability scores. All the vulnerability scores of households were standardized to get uniform scale (ranges from 1–100). The results indicated that most (63.06%) of the slum households while only 20.02% non-slum households were highly vulnerable to flood. The present study also identified and evaluated the responsible factors that create people's vulnerability to flooding hazards in Dhaka megacity. The key feature of this paper is to provide a unique method and model to calculate numeric household vulnerability to flooding hazards, and this practical approach is useful to quantify hazard-induced vulnerabilities not only for Dhaka but also for other cities especially for developing countries.

Md. E. Huq (✉) · A. Javed · Z. Shao
State Key Laboratory of Information Engineering in Surveying, Mapping and Remote Sensing, Wuhan University, Wuhan, PR China
e-mail: enamul_huq@whu.edu.cn; shaozhenfeng@whu.edu.cn

A. Z. M. Shoeb
Department of Geography and Environmental Studies, University of Rajshahi, Rajshahi, Bangladesh

M. A. Hossain
Department of Geography and Environment, Jagannath University, Dhaka, Bangladesh

M. S. Sarven
College of Plant Science and Technology, Huazhong Agricultural University, Wuhan, Hubei, PR China

© Springer Nature Switzerland AG 2020
X. Yuan, M. Elhoseny (eds.), *Urban Intelligence and Applications*, Studies in Distributed Intelligence, https://doi.org/10.1007/978-3-030-45099-1_16

Keywords Flood hazard · Household vulnerability · Vulnerability index ·
AHP · Slum · Dhaka megacity

1 Introduction

The consequences of natural hazards are likely to increase the vulnerability to
general populations [1]. The physical damages from natural disasters have increased
significantly from the last decades. Floods are accounting for 47% of all weather-
related disasters from 1995 to 2015, affecting 2.3 billion people and killing 157,000
[2]. 80% of the total population affected by riverine flooding each year is relegated
to 15 least developed or developing countries in Africa, Asia, and America [3,
4]. Unfortunately, extreme flood events have become more frequent and harmful
in Europe for the last 25 years regarding the duration and extension of affected
areas [5]. Among the other countries, Bangladesh is one of the most flood-prone
countries in the world. Flood has been a regular phenomenon not only in rural
areas but also in the urbanized Dhaka and its adjoining areas [6–10]. According
to [11], about 20 million people in Bangladesh are living with the risk of flood.
Particularly, the capital city Dhaka suffers a lot from flood hazards [12, 13]. The
fact is that poor people tend to live in high-risk areas that are responsible to create
vulnerability to epidemic [14, 15]. Therefore, they are most vulnerable to flood-
related disasters. On the other hand, limited resources, improper planning, fast
urban development, and increasing population are raising the vulnerability of urban
people to cope with disasters [16, 17]. The objective of this paper is to develop a
vulnerability index (VI) that is capable of measuring households' vulnerability to
flooding hazards and evaluating the relative magnitude of flood vulnerability. The
proposed methodological framework in determining VI is expected to make a bridge
between qualitative and quantitative vulnerability assessments.

2 Materials and Methods

2.1 Study Area and Data Collection

The study was carried out in Dhaka megacity, the capital of Bangladesh. It is the
home of more than 15 million people [18]. Dhaka is known as a city of slums. It was
reported by the Center for Urban Studies (CUS) (2006) that there are 4966 clusters
of slum and squatter in Dhaka city that were 2156 in 1991 and 3007 in 1996. The
specific study area situated between latitudes $23°39'$ and $23°54'$N and longitudes
$90°20'$ and $90°28'$E (Fig. 1) was chosen to evaluate the flood vulnerability at the
household level.

The household-level survey was conducted in the flood-affected settlement
(Fig. 1). Approximately 1900 households reside in the study area; among them

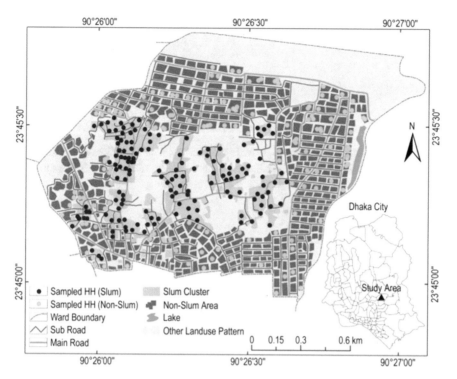

Fig. 1 Location and sampling households of the study area (modified from the Center for Urban Studies (CUS) (2006))

300 (150 from non-slums and 150 from slums) have been interviewed through a questionnaire survey. Based on the social, economic, structural, institutional, environmental, and demographic factors a structured questionnaire was used to collect primary data for calculating households' vulnerability to flooding hazards.

2.2 Computing of Vulnerability Index (VI)

This study endeavors to integrate the best method in the course of developing VI. For assigning weight values to the variables and indicators, Analytic Hierarchy Process (AHP) is used [19]. To measure the relative importance of the indicators, a pairwise comparison method is followed. The numeric values from 1 to 9 are assigned based on the importance of the variables or indicators [20, 21]. From the overall performance matrix (OPM) (Table 1), the weight of the criterion was calculated through normalizing the cells of columns (divide the cell value with the sum of a column). However, the reliability of comparisons was calculated by computing the consistency ratio (CR).

		RN	HT	TS	S	Weights
Table 1 Overall preference matrix and weights	RN	1	1/3	5	1	0.232
	HT	3	1	5	1	0.402
	TS	1/5	1/5	1	1/5	0.061
	S	1	1	5	1	0.305
	CR = 0.073					

The CR maintains balance to provide weights upon others. The acceptable ratio of CR is <0.10. However, the VI has considered in the range of 1–100 scale. The higher value represents the high vulnerability to flooding.

3 Results and Discussion

Based on the previous studies and field observation the current study selects five factors (social, economic, structural, environmental, and institutional) of human vulnerability to evaluate the flood vulnerability of urban people.

3.1 Degree of Structural Vulnerability

Structural vulnerability is another influential factor in flood hazards. It can magnify the intensity of flood hazards [22, 23]. Under structural factors, four variables (housing type, emergency shelter during the flood event, road network, and transport system for evacuation) are included. The result exhibits that most of the households in the slum area were under high vulnerability (82%) with respect to structural variables and 18% were moderately vulnerable. So, most probably the slum inhabitants will seek or expect emergency shelter during natural hazards such as flooding. Surprisingly in consideration of structural vulnerability almost all (95.3%) non-slum households were moderately vulnerable, few were highly vulnerable (4.7%), and there was none in low vulnerable category (Table 2).

3.2 Degree of Social Vulnerability

In this study, the social factor includes education, occupation, type of household head, family size, disabled person, preparation, and social network. The study found that the aggregated social vulnerability for slum households/inhabitants leads by moderate vulnerability (67.3%). 16% and 16.7% households are low and highly vulnerable. The picture is different in the non-slum area where a large number of households (78.7%) are in low vulnerable category, some (18.7) are moderate, and

Table 2 Degree of structural, social, economic, environmental, and institutional vulnerability

Levels of vulnerability	Structural vulnerability		Social vulnerability		Economic vulnerability		Environmental vulnerability		Institutional vulnerability	
	Slum (%)	Non-slum (%)	Slum (%)	Non-slum (%)	Slum (%)	Non-slum (%)	Slum (%)	Non-slum (%)	Slum (%)	Non-slum (%)
Low	00	00	16.0	78.7	00	14.0	4.0	21.3	15.3	22.7
Moderate	18.0	95.3	67.3	18.7	16.0	59.3	32.7	78.0	15.3	12.0
High	82.0	4.7	16.7	2.7	84.0	26.7	63.3	0.7	69.3	65.3
Total	100	100	100	100	100	100	100	100	100	100

very few (2.7) are in the highly vulnerable category. It indicates that the slum-people are neither at low risk nor at high, they are rather at moderate risk if flood would occur.

3.3 Degree of Economic Vulnerability

Here, income, land ownership, savings, insurance, loan status, and ownership of vehicles were considered to estimate the economic vulnerability of a household. It is a critical factor to get an idea of economic strength to tackle the hardship of potential flood hazards. The present study finds that the majority of households (84%) of the slum are economically highly vulnerable and rest (16%) is moderate. In non-slum, the scenario is almost reverse because more than half (59.3%) of the households are moderately and 26.7% are highly vulnerable to potential floods (Table 2). Remaining 14% households fall into low vulnerable categories due to their strong economic conditions. It is observed that the vital strength for the livelihood is quite fragile for the slum-people to combat disasters.

3.4 Degree of Institutional Vulnerability

In this factor, two variables were considered (aid and forecasting). The empirical result represents that, the majority (69.3% and 65.3%) of slum and non-slum households were highly vulnerable owing to institutional support. 15.3% households from slum and 12% from non-slum were moderately and only 15.3% and 22.7% of households from both slum and non-slum were vulnerable in terms of institutional vulnerability (Table 2). Interestingly, the trend of the institutional vulnerability of slum and non-slum households is very close to each other. So, it means that both slum and non-slum inhabitants are careless and unaware of institutional background. They count this factor as almost useless to their daily life activities. Rather it is observed that they count the economic factor most.

3.5 Level of Environmental Vulnerability

The environmental sphere cannot be separated from social and economic spheres because of mutual interaction between social and natural environment [22, 24].

In this study, only two variables (toilet facility and accessibility of drinking water) were incorporated to evaluate the environmental vulnerability for both slum and non-slum inhabitants. After measuring these two variables, it was found that among slum inhabitants 63.3% were environmentally at high risk, 32.7% were at moderate, and only 4.0% were at low environmental risk. In contrast, this scenario is

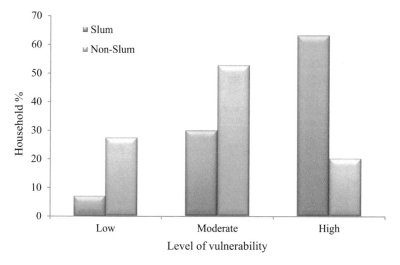

Fig. 2 The overall household vulnerability of urban inhabitants to flood

quite reverse for the non-slum inhabitants. It was observed that only 0.7% non-slum habitats were at high risk, 21.3% were at moderate, and 78% were environmentally at low vulnerable category (Table 2).

4   Overall Vulnerability

The overall human vulnerability to flood had been developed by aggregating [15, 25] of all vulnerability factors. The result presents that the majority (63.06%) of slum households and 20.02% of non-slum were highly vulnerable to floods in the study area. In addition, 29.88% and 7.06% slum households were moderately and lowly vulnerable. On the other hand, in non-slum, 52.66% households were moderately, and 27.34% households were lowly vulnerable to future floods (Fig. 2).

5   Conclusion

This study assessed the vulnerability status of slum and non-slum inhabitants in Dhaka city with the help of longitudinal analysis on households' responses towards flood impacts. The results show that the newly developed VI is a simple but very effective approach to show the state of household vulnerability to flooding hazards. The analytical result demonstrated that most of the households (63.06%) of the slum were highly vulnerable to flood hazard. Whereas, 52.66% households of non-slum were moderately vulnerable for future flood events. The study also found that the social network of slum-people is stronger than that of non-slum people. The uniqueness of this study is two folds: firstly, primary data is used to develop VI; secondly, AHP is exploited to assign the weights of each variable and indicators. This method has been applied in Dhaka city but it is possible to be extended for other cities in developing countries where socio-economic infrastructures are similar.

Acknowledgment This work was supported in part by the National Key R & D plan on strategic international scientific and technological innovation cooperation special project under Grant 2016YFE0202300, the National Natural Science Foundation of China under Grants 61671332, 41771452, and 41771454, the Natural Science Fund of Hubei Province in China under Grant 2018CFA007.

References

1. M.N. Halgamuge, A. Nirmalathas, Analysis of large flood events: based on flood data during 1985–2016 in Australia and India. Int. J. Disaster Risk Reduct. **24**, 1–11 (2017)
2. A. Atreya, J. Czajkowski, W. Botzen, G. Bustamante, K. Campbell, B. Collier, F. Ianni, H. Kunreuther, E. Michel-Kerjan, M. Montgomery, Adoption of flood preparedness actions: a household level study in rural communities in Tabasco, Mexico. Int. J. Disaster Risk Reduct. **24**, 428–438 (2017)
3. S.D. Brody, W.E. Highfield, R. Blessing, T. Makino, C.C. Shepard, Evaluating the effects of open space configurations in reducing flood damage along the Gulf of Mexico coast. Landsc. Urban Plan. **167**, 225–231 (2017)
4. J.C.V. de Leon, Vulnerability assessment: the sectoral approach. Meas. Vulnerability Nat. Hazards **2007**, 300–315 (2007)
5. W.B.M. Ten Brinke, J. Knoop, H. Muilwijk, W. Ligtvoet, Social disruption by flooding, a European perspective. Int. J. Disaster Risk Reduct. **21**, 312–322 (2017)
6. A. Masuya, A. Dewan, R.J. Corner, Population evacuation: evaluating spatial distribution of flood shelters and vulnerable residential units in Dhaka with geographic information systems. Nat. Hazards **78**, 1859–1882 (2015)
7. A.S. Chen, M.J. Hammond, S. Djordjević, D. Butler, D.M. Khan, W. Veerbeek, From hazard to impact: flood damage assessment tools for mega cities. Nat. Hazards **82**, 857–890 (2016)
8. A.K. Gain, V. Mojtahed, C. Biscaro, S. Balbi, C. Giupponi, An integrated approach of flood risk assessment in the eastern part of Dhaka City. Nat. Hazards **79**, 1499–1530 (2015)
9. M.B. Sciance, S.L. Nooner, Decadal flood trends in Bangladesh from extensive hydrographic data. Nat. Hazards **90**, 115–135 (2018)
10. K.R. Rahaman, Social capital and good governance—a nexus for disaster management: lessons learned from Bangladesh, in *Living Under the Threat of Earthquakes: Short and Long-Term Management of Earthquake Risks and Damage Prevention in Nepal*, ed. by J. H. Kruhl, R. Adhikari, U. E. Dorka, (Springer, Cham, 2018), pp. 211–228
11. C. EM-DAT, The OFDA/CRED international disaster database, Université catholique (2014)
12. A.K. Azad, K.M. Hossain, M. Nasreen, Flood-induced vulnerabilities and problems encountered by women in northern Bangladesh. Int. J. Disaster Risk Sci. **4**, 190–199 (2013)
13. M.E. Huq, *Flood Hazard, Vulnerability and Adaptation of Slum Dwellers in Dhaka* (Lambert Academic Publishing, Saarbrücken, 2013)
14. S. Schneiderbauer, D. Ehrlich, Social levels and hazard (in) dependence in determining vulnerability. Meas. Vulnerability Nat. Hazards **2006**, 78–102 (2006)
15. C. Sebald, Towards an integrated flood vulnerability index: a flood vulnerability assessment, Master of Science (MSc) (2010)

16. L.-F. Melgarejo, T. Lakes, Urban adaptation planning and climate-related disasters: an integrated assessment of public infrastructure serving as temporary shelter during river floods in Colombia. Int. J. Disaster Risk Reduct. **9**, 147–158 (2014)

17. M.E. Huq, Analyzing vulnerability to flood hazard of urban people: evidences from Dhaka megacity, Bangladesh. Int. J. Earth Sci. Eng. **10**, 585–594 (2017)

18. K.M. Bahauddin, M.M. Rahman, F. Ahmed, Towards urban city with sustainable buildings a model for Dhaka City, Bangladesh. Environ. Urban. ASIA **5**, 119–130 (2014)

19. T.L. Saaty, How to make a decision: the analytic hierarchy process. Eur. J. Oper. Res. **48**, 9–26 (1990)

20. G. Coyle, The analytic hierarchy process (AHP). Practical strategy: structured tools and techniques (2004)

21. M.R. Rahman, S. Saha, Flood hazard zonation–a GIS aided multi-criteria evaluation (MCE) approach with remotely sensed data. Int. J. Geoinform. **3**, 25–37 (2007)

22. R.K. Samanta, G.S. Bhunia, P.K. Shit, H.R. Pourghasemi, Flood susceptibility mapping using geospatial frequency ratio technique: a case study of Subarnarekha River Basin, India. Model. Earth Syst. Environ. **2018**, 1–14 (2018)

23. M.E. Huq, M.A. Hossain, Vulnerability framework for flood disaster management. J. Geo-Environ. **11**, 51–67 (2015)

24. O.D. Cardona, Disaster risk and vulnerability: concepts and measurement of human and environmental insecurity, in *Coping with Global Environmental Change, Disasters and Security: Threats, Challenges, Vulnerabilities and Risks*, ed. by H. G. Brauch, Ú. Oswald Spring, C. Mesjasz, J. Grin, P. Kameri-Mbote, B. Chourou, P. Dunay, J. Birkmann, (Springer, Berlin, 2011), pp. 107–121

25. M.E. Huq, M.A. Hossain, Flood hazard and vulnerability of slum dwellers in Dhaka. Stamford J. Environ. Human Habitat **1**, 36–47 (2012)

Extracting Keyphrases from News Articles Using Crowdsourcing

Qingren Wang, Jinqin Zhong, Lichuan Gu, Kai Yang, and Victor S. Sheng

Abstract Keyphrase extraction is a very important task in text mining. However, keyphrase extraction of news articles cannot be addressed by existing machine-based approaches effectively because of various reasons. This paper employs crowdsourcing for keyphrase extraction of news articles. We first design a proper crowdsourcing mechanism to extract keyphrases from news articles and then adapt three truth inference algorithms (namely IMLK, IMLK-I, and IMLK-ED) for integrating multiple lists of keyphrases provided by workers. The experiments show that crowdsourcing can significantly improve the performance of the machine-based approach (i.e., KeyRank).

Keywords Keyphrase extraction · Crowdsourcing · Ground truth inference · Grade calculation

Q. Wang
School of Computer Science and Technology, Anhui University, Hefei, China
e-mail: wqr@ahu.edu.cn

J. Zhong
International Business School, Anhui University, Hefei, China

L. Gu
School of Information and Computer, Anhui Agricultural University, Hefei, China
e-mail: gulichuan@ahau.edu.cn

K. Yang
Department of Computer Science, Tongji University, Shanghai, China
e-mail: kaiyang@tongji.edu.cn

V. S. Sheng (✉)
Department of Computer Science, University of Central Arkansas, Conway, AR, USA
e-mail: ssheng@uca.edu

X. Yuan, M. Elhoseny (eds.), *Urban Intelligence and Applications*, Studies in Distributed Intelligence, https://doi.org/10.1007/978-3-030-45099-1_17

217

1 Introduction

A keyphrase in a document is a list of words with sequential order that captures the main point expressed by the document [1, 2], which means keyphrases can greatly help users to understand the topics discussed in the document. Keyphrase extraction has been successfully used in automatic indexing, topic extraction, text summarization and categorization, etc. Because of keyphrases' advantages and importance, many studies have been conducted for extracting keyphrases with high quality from documents based on machine learning. However, keyphrase extraction served in personalized news recommendations cannot be effectively addressed by existing machine-based approaches. The functionality of personalized news recommendations is to recommend accurate and useful news articles with timeliness, suddenness, and novelty to users. However, such news articles usually are short and include new emerging key entities with low frequencies, which results in machine-based approaches lose their abilities since (1) existing machine-based approaches are frequency-based, which cannot extract important entities because of low frequencies; (2) existing knowledge bases and ontologies may become out-of-date and cannot quickly be updated to new ones because of timeliness, suddenness, and novelty. To sum up, keyphrase extraction served in personalized news recommendation is an intelligent and computer-hard task, so that it is natural to think of employing the intelligence of human beings.

Incontestable, hiring domain experts to extract keyphrases can achieve high accuracy. However, it spends a long time as well as high resources. Researches [3–5] demonstrated that crowdsourcing brings machine learning as well as its related fields excellent opportunities. Since crowdsourcing can easily access the crowd because of public platforms [6], such as AMT [7] and CrowdFlower [8], it can efficiently deal with intelligent and computer-hard tasks by employing thousands of workers. Intuitively, news articles recommended to general public are normally short and easy to understand, which do not require workers to have a high level of literacy. Therefore, this paper employs crowdsourcing to extract keyphrases, and name it *crowdsourced keyphrase extraction*. Accordingly, tasks published by crowdsourced keyphrase extraction are called *KE-HITs* (which will be introduced in Sect. 3.1). This paper applies crowdsourcing to extract keyphrases from news articles and investigate whether crowdsourced keyphrase extraction can achieve a more effective performance than existing machine-based approaches.

Since keyphrase extraction is a complex process that includes extracting and ranking, a single KE-HIT includes three task types: *multiple-choice*, *fill-in-blank*, and *rating*. These three task types are widely utilized in the public crowdsourcing platforms, e.g., AMT and CrowdFlower, where *Multiple-choice* and *fill-in-blank* are for the crowd to extract proper keyphrases, and rating for the crowd is used to provide importance rankings for the extracted keyphrases. This is different from most crowdsourced tasks published on AMT, which usually are simple and have only one task type. A single KE-HIT is more complicated and full of challenges, especially for following keyphrase integration. Besides, there are three important problems in crowdsourced tasks which also required to be balanced in crowdsourced keyphrase extraction—*quality control*, *cost control*, and *latency control* [6]. Quality control focuses on generating truth keyphrases with high quality for a task, cost control aims to reduce the costs in terms of labor and money while keeping high-quality truth, and latency control studies how to cut down the time [9]. In this paper we utilize the following methods to handle the trade-off among quality control, cost control, and latency control.

- We adopt a pruning-based technique [6] to prune the keyphrases extracted by a machine-based approach. The pruning-based technique can efficiently reduce human cost and workers' latency (which means it can reduce the time cost). Besides, we add an additional option in our crowdsourcing mechanism, which asks the crowd to supplement proper keyphrases that are lost because of various reasons.
- Since time constraints are important to control latency [9], we set a time constraint for each single KE-HIT. In addition, we ask the crowd to select an importance ranking for each keyphrase, instead of directly asking the crowd to sort keyphrases.
- In order to conquer the possible low quality of some crowdsourcing workers for keyphrase extraction, our crowdsourcing mechanism allows multiple workers [4] to extract keyphrases from a news article.

The remainder of the paper is organized as follows. After introducing the related work in Sect. 2, the details of crowdsourced keyphrase extraction will be discussed in Sect. 3, the experimental results will be reported in Sect. 4, and then we will reach a conclusion in Sect. 5.

2 Related Work

Most original works extracting keyphrases treat a single word with high frequency or a list of frequent contiguous words as a keyphrase, such as KEA [11]. Yet these single or contiguous words do not always capture the main points expressed in a document. Studies [12] demonstrated that semantic relations in context can greatly help accelerate the quality of extracted keyphrases. Therefore, some researches [13, 14] imported knowledge bases or ontologies to obtain semantic relations to accelerate the accuracy of extracting keyphrases. Fu et al. developed several keyphrase semantic extension ranked schemes based on conceptual graphs under different semantic search scenarios [13, 14]. Some studies [15, 16] applied sequential pattern mining with wildcards to extract keyphrases, as wildcards can afford flexible gap constraints for mining keyphrase candidates capturing semantic relations in the context in a document [17]. KeyRank [10], which only scans a document once to find frequent words and their corresponding positions that appear in the document, is an effective algorithm extracting keyphrases based on sequential pattern mining and entropy. However, it is also frequency-based algorithm that may lose important entities with low frequencies. To sum up, keyphrase extraction is a computer-hard task which cannot effectively be addressed by existing machine-based approaches.

Studies [3–5] revealed that crowdsourcing brings machine learning as well as its related fields great opportunities. With the development of public crowdsourcing platforms, e.g., AMT [7] and CrowdFlower [8], crowdsourcing has taken off in a wide range of applications. As a crowd-powered database system, CDB [18] supports crowd-based query optimization focusing on selection and join. The ESP game [19] first employs the crowd to recognize objects in images and then collects the labels for object recognition.

Since some individuals in the crowd may yield relatively low-quality answers or even noise, many studies focus on how to infer truths based on workers' answers [6]. Zheng et al. [20] not only provided a particular survey on truth inference on crowdsourcing but also analyzed 17 existing approaches in detail. However, these truth inference algorithms cannot be utilized in crowdsourced keyphrase extraction for integrating and re-grading the multiple lists of answers provided by workers. One of our preliminary works [21] views truth inference as a procedure of integrating and re-grading answers and proposed three novel truth inference algorithms, namely IMLK, IMLK-ED, and IMLK-I. They first figure out the grade of each answer based on its own different rankings provided by workers and then aggregates all answers with grades together to generate a new integrated answer list with orderliness. Finally, they output the well-organized integrated answer list, where users could select top-n ones as their required answers.

3 Crowdsourced Keyphrase Extraction

3.1 KE-HIT

The experiments of crowdsourced keyphrase extraction are conducted on a popular crowdsourcing platform, namely Amazon Mechanical Turk (AMT in short). AMT supports crowdsourced execution of Human Intelligence Tasks (HITs in short) [19]. The tasks published in our studies are called *Keyphrase Extraction Human Intelligence Tasks* (KE-HITs). That is, the structure of a single KE-HIT is inherited from a HIT executed by AMT essentially. Since keyphrase extraction is a complex process including extracting and ranking, and a single news article is treated as the smallest unit during the execution, each single news article corresponds to a KE-HIT that fits the structure and content of the news article. Besides, we utilize a pruning-based technique that employs top-n keyphrases extracted by a machine-

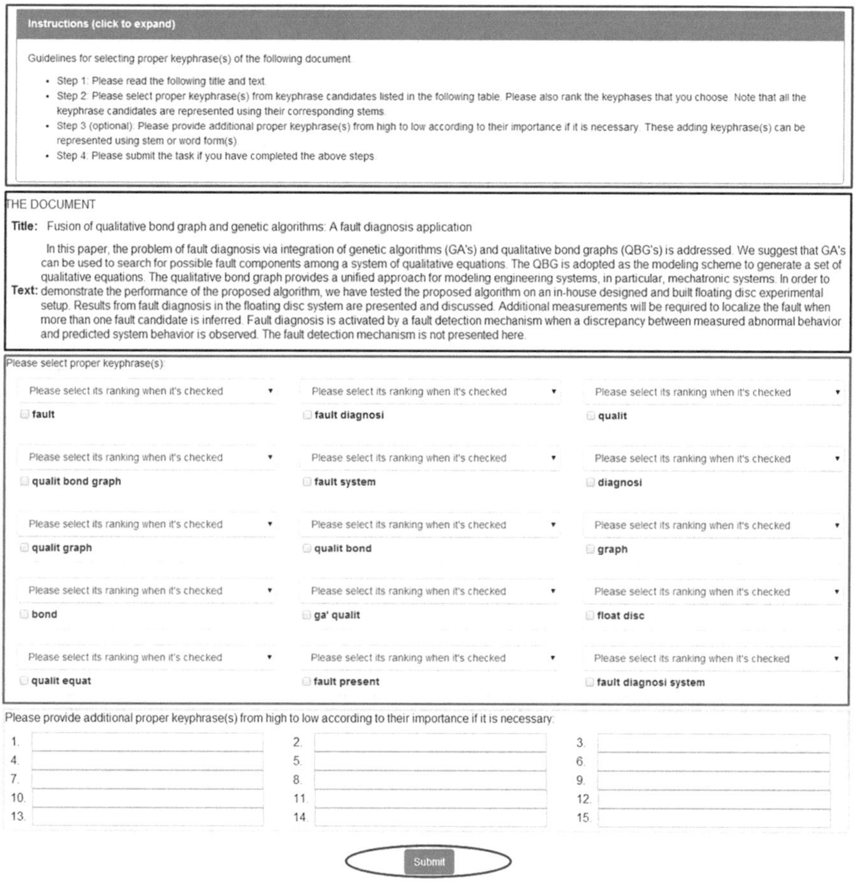

Fig. 1 The user interface of a single KE-HIT

based approach to balance problems, quality, cost, and latency in KE-HITs. At last, a single KE-HIT is composed of five parts: instruction, document, keyphrase selection, keyphrase offering, and operation (see Fig. 1). The instruction part guides workers to complete tasks easier and faster (see the blue rectangle in Fig. 1). The document part displays a news article's title and content (see the black rectangle in Fig. 1). The operation part helps workers submit the KE-HIT (see the blue ellipse in Fig. 1).

The keyphrase selection part shows workers candidates extracted by a machine-based approach (see the red rectangle in Fig. 1). Note that this part only holds 15 candidates at most in this paper. If a document has more than 15 candidates, it only shows the top-15 ones with the highest scores. The keyphrase selection part has two task types as follows.

- *Multiple-choice.* A worker directly selects the proper keyphrase(s) from this part.
- *Rating.* Once a candidate is selected as a proper keyphrase, the worker needs to assign an importance ranking to it. The importance of ranking denotes how important a keyphrase is to the news article. It varies from 1 to 15, where 1 denotes the most important and 15 denotes the least important.

Some proper keyphrase(s) may not be listed in the keyphrase selection part as they have lower scores assigned by machine-based approaches, or may not have been extracted from the document yet because of low frequencies. That is why a KE-HIT has the keyphrase offering part, which lets workers supplement more proper keyphrase(s) as well as corresponding importance rankings (see the yellow rectangle in Fig. 1). Note that the keyphrase offering is optional.

3.2 Keyphrase Lists Generation

The procedure for generating keyphrase lists is divided into two progressive parts: *keyphrase voting* and *ranking calculation*. *Keyphrase voting* gathers applicable keyphrases from a news article by employing workers to vote for candidates that are extracted by a machine-based approach or themselves. The steps of *keyphrase voting* are as follows.

- Top-n candidates, which are chosen from a keyphrase set produced by a machine-based approach, are shown to workers.
- Asking workers to select the proper keyphrase(s) from the candidates, and to assign importance ranking(s).
- Workers can provide additional proper keyphrases as well as their importance rankings if they are willing to.
- Workers submit completed tasks.

Ranking calculation starts after finishing *keyphrase voting*. Before then, some pre-processing approaches (i.e., stop-word and punctuation-mark removing, stemming) are utilized to standardize keyphrases afforded by workers. Note that this step employs a man-made parameter named *ranking score* to assist in truth inference. The steps of *ranking calculation* are as follows.

- The aggregated keyphrases are divided into different *k*eyphrase *l*ists (KLs) based on their suppliers, which means each KL corresponds to an individual worker. Each KL is well-organized because keyphrases in each KL are sorted by their own importance rankings (provided by workers) from high to low temporarily. Besides, the union (denoted as U) of different keyphrases afforded by workers is obtained through aggregating keyphrases from all KLs.
- Calculating the *ranking score* of each keyphrase list by list. A keyphrase in each KL has its own ranking score, which is defined as follows. Note that these ranking scores are independent of each other.

$$RS_{ij} = \#DIF - TRN_{ij} + 1 \qquad (1)$$

where RS_{ij} represents the ranking score of the ith keyphrase k in the jth KL, TRN_{ij} represents the importance rankings of the ith k in the jth KL, and #DIF represents the length of U. The ranking score of k in a KL is set to zero if it does not show up.

3.3 Algorithms for Inferring Ground Truth

In some sense, keyphrase candidates extracted from a document could be treated as correct keyphrases of this document, since each one definitely delivers a certain point of the document. The only difference between them is that they have different grades of delivering points. Therefore, the probability (used in supervised-learning-based approaches) of a candidate to be a keyphrase is the grade of the keyphrase, and the ranking score (used in approaches based on unsupervised learning) of a keyphrase candidate is its own grade. Besides, for a keyphrase candidate which has not been extracted, it is viewed as a keyphrase with grade zero.

Therefore, we view truth keyphrase inference as a procedure of first-integrating last-grading keyphrases, i.e., all keyphrases from different KLs are first aggregated together, and then each keyphrase's grade is calculated according to its own different ranking scores. After then, the aggregated keyphrases are re-ranked based on their grades from high to low. Finally, a new well-organized *integrated keyphrase list* (IKL) is inferred as the truth list, where users can select top-n ones as their demands.

In this paper, we adapt algorithms IMLK, IMLK-I, and IMLK-ED proposed in [21] to execute the procedure of first-integrating last-grading keyphrases, since they are suitable for integrating and re-grading the true keyphrase list from the plenty of lists of keyphrases. IMLK considers that workers have the same quality while they are dealing with the same task. It utilizes (2) to re-calculate the grade of a keyphrase.

$$G_{ij} = \frac{\sum_{j=1}^{m} RS_{ij}^{2}}{m} \qquad (2)$$

where G_{ij} represents the grade of the ith keyphrase in the union U, RS_{ij} represents the ranking score of the ith keyphrase in the jth KL, and m denotes the number of workers hired.

However, in reality, workers working on crowdsourcing platforms do have different qualities because of their personalities, expertise, and biases [6, 9], which conflicts with the premise of IMLK—workers have the same quality of dealing with the same task. Although IMLK-I and IMLK-ED are proposed to solve the weakness

[21], they still ignore three inherent properties of a keyphrase [10]: meaningfulness, uncertainty, and uselessness.

The adapted IMLK-ED in the paper utilizes entropy [22] and follows the equations in [10] to integrated estimate the inherent properties of a keyphrase capturing the main point. Firstly, the probability of a keyphrase k as an independent keyphrase (represented as P_i) is employed to measure k's meaningfulness.

$$P_i = \begin{cases} \#NIC/\#TN, & 0 < NIC < TN \\ 0, & NIC = 0 \end{cases} \tag{3}$$

where #NIC represents the number of times of k as an independent keyphrase appearing in the corpus, and #TN represents the total quantity of keyphrases the corpus has.

The probability of k as a sub-keyphrase (denoted as P_s) is utilized to measure k's uncertainty.

$$P_s = \begin{cases} \#NSC/\#TN, & 0 < NSC < TN \\ 0, & NSC = 0 \end{cases} \tag{4}$$

where #NSC represents the number of times of k as a sub-keyphrase appearing in the corpus.

The uselessness of k capturing the main point is described as the probability of other situations (denoted as P_o).

$$P_o = 1 - P_i - P_s \tag{5}$$

In summary, the entropy of k (denoted as $H(k)$) is defined by (6). If the situation of "$P_s = 0$" occurs, (7) is used to figure out $H(k)$.

$$H(k) = P_i \, \log\,(1/P_i) + P_s \, \log\,(1/P_s) + P_o \, \log\,(1/P_o) \tag{6}$$

$$H(k) = P_i \, \log\,(1/P_i) + P_o \, \log\,(1/P_i) \tag{7}$$

In order to evaluate the quality of a worker, the adapted IMLK-ED first calculates each KL's distance (denoted as Dis). The Dis of a KL denotes how many operations are required to change this KL to IKL. After all the distances of KLs are obtained, the total distance *SumDis* is figured out too.

$$Q_j = 1 - \left(Dis_j / SumDis\right) \tag{8}$$

where Q_j represents the quality of a worker providing the jth KL, and Dis_j represents the edit distance of changing the jth KL to IKL. The adapted IMLK-ED has five steps as follows.

1. It calculates the entropy of each keyphrase.
2. It employs the IKL generated by IMLK as the initial IKL at the first iteration, then it adopts the IKL generated by the previous iteration as the initial IKL to start a new iteration.
3. It calculates the distances of KLs as well as SumDis.
4. It evaluates the quality of workers using (8).
5. It re-calculates the grade of each keyphrase using (9).

$$G_{ij} = \frac{\sum_{j=1}^{m} Q_j \cdot H(k)_{ij} \cdot RS_{ij}^{2}}{m} \tag{9}$$

where $H(k)_{ij}$ represents the entropy of the ith keyphrase in the jth KL. Then it re-integrates all keyphrases with new grades together to form a new ordered IKL. Finally, it goes back to Step 2 and iteratively updates its IKL until convergence occurs.

Since the adapted IMLK-I in the paper also utilizes entropy for measuring the inherent properties of a keyphrase, we will not go into that again.

4 Experiments

Since KeyRank performs very well [10], we employed KeyRank as the baseline as well as keyphrases extracted by KeyRank for our pruning-based technique. Because abstracts in INSPEC [23] are short and easy to understand, which are similar to news articles utilized in the personalized news recommendation, we chose 50 documents from the 500 test abstracts of INSPEC, where KeyRank performs the best in extracting high-quality keyphrases, as the original data for our crowdsourcing experiments. As mentioned earlier, every single document corresponds to a single KE-HIT. Thus, we have 50 KE-HITs.

For the crowdsourcing experiment conducted in this paper, each KE-HIT requires 10 lists of answers from 10 different workers working on AMT, which means each KE-HIT needs to be published ten times. Thus, the crowdsourcing experiment has 500 published KE-HITs on AMT at last. Each KE-HIT costs 5 cents, and the crowdsourcing experiment costs 35 dollars totally (25 dollars for paying 500 KE-HITs and 10 for offering management of AMT). Besides, the precision (represented as P), recall (represented as R), and F_1 score are used to measure the performance of crowdsourcing experiment.

$$P = \#correct/\#provided \tag{10}$$

$$R = \#correct/\#labeled \tag{11}$$

$$F_1 = 2 \times P \times R/(P + R) \tag{12}$$

where #correct represents the number of correctly extracted keyphrases, #provided represents the number of keyphrases provided by the crowd, and #labeled represents the number of correct keyphrases offered by authors or experts. The number of extracted keyphrases of each document for performance metrics in the crowdsourcing experiments varies from 3 to 5.

After the 10 lists of answers from 10 different workers for each KE-HIT are obtained, algorithms IMLK, IMLK-I, and IMLK-ED are applied for integrating the 10 lists of answers. The experimental results of the three algorithms are compared with those of KeyRank in terms of precisions, recalls, and F_1 scores. In addition, in order to clearly show their performance, the comparison results among four algorithms (i.e., IMLK, IMLK-I, IMLK-ED, and KeyRank) are divided into four different categories. That is, category top-3, category top-4, category top-5, and category average. For example, category top-5 is named because the quantity of extracted keyphrases is 5 when it reports the comparison results among the four algorithms in terms of P, R, and F_1 score, respectively. Category average exhibits the average performance of four algorithms in categories top-3, top-4, and top-5. In addition, the relationships between the quantity of workers and the performance of experiment are also studied by respectively building another eight comparisons in terms of workers' quantity, and the corresponding results are also reported in terms of precisions, recalls, and F_1 scores. The numbers of workers are respectively set to 3, 4, 5, 6, 7, 8, 9, and 10. In order to eliminate the influence generated by the selection order of workers, (1) each time the corresponding quantity of selecting answer lists from the total 10 answer lists is randomly chosen (For example, when the quantity of workers is 6, we randomly choose 6 answer lists from the total 10 answer lists); and (2) when the comparisons among IMLK, IMLK-I, and IMLK-ED are built under a certain quantity of selecting answer lists, each algorithm is run ten times on each document. All experimental comparisons among IMLK, IMLK-I, IMLK-ED, and KeyRank are respectively shown in Figs. 2, 3, and 4.

From Figs. 2, 3, and 4, we notice that IMLK significantly performs better than KeyRank in terms of P, R, and F_1 score in categories top-3, top-4, and top-5, except the experimental comparison in terms of R in category top-5 when the quantity of workers is 3 (see the left part in Fig. 3c). We also notice that both IMLK-I and IMLK-ED significantly outperform KeyRank and IMLK in terms of P, R, and F_1 score in categories top-3, top-4, and top-5. Besides, except the comparisonsin terms of P, R, and F_1 score in categories top-3, top-4, and top-

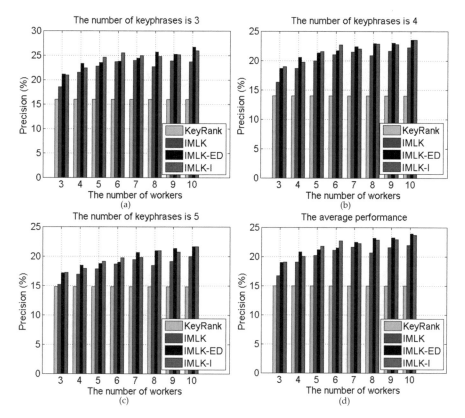

Fig. 2 The precision comparisons among four algorithms in (**a**) Category top-3, (**b**) category top-4, (**c**) category top-5, and (**d**) category average

5 when the numbers of workers are 5, 6, and 7 (the situation that the number of workers is 7 only occurs in category top-3), IMLK-ED always outperforms IMLK-I. Moreover, the comparisons in category average (see Figs. 2d, 3d, and 4d) show that the performance of IMLK, IMLK-I, and IMLK-ED has a rising trend while the quantity of workers is increasing. To sum up, we conclude that: (1) IMLK is simple but is more efficient than machine-based approaches; (2) both IMLK-I and IMLK-ED can improve the performance of IMLK, and IMLK-ED outperforms IMLK-I slightly; (3) the number of workers impacts the performance of algorithms IMLK, IMLK-I, IMLK-ED; and (4) utilizing crowdsourcing is a feasible and effective approach to extract proper keyphrases from news articles.

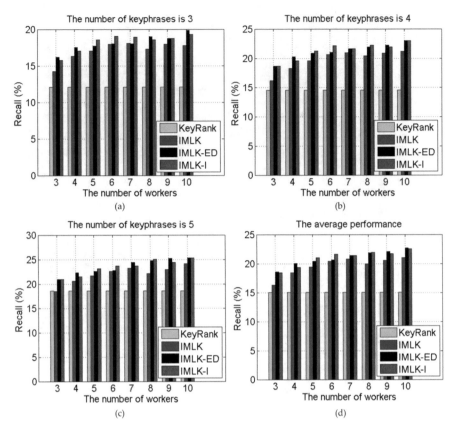

Fig. 3 The recall comparisons among four algorithms in (**a**) Category top-3, (**b**) category top-4, (**c**) category top-5, and (**d**) category average

5 Conclusion

We imported crowdsourcing into the keyphrase extraction field, designed novel crowdsourcing mechanisms to collect multiple lists of answers from different workers, and then adapted algorithms IMLK, IMLK-I, and IMLK-ED for integrating and re-grading the multiple lists of answers from different workers. The experimental results demonstrated that crowdsourced keyphrase extraction can achieve a more effective performance than KeyRank does.

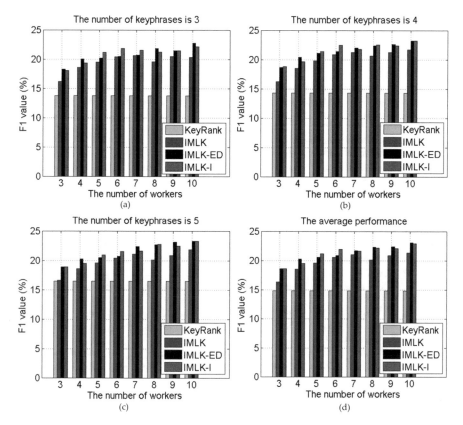

Fig. 4 The F_1 score comparisons among four algorithms in (**a**) Category top-3, (**b**) category top-4, (**c**) category top-5, and (**d**) category average

Acknowledgment This work is partially supported by the US National Science Foundation under grant IIS-1115417, the National Natural Science Foundation of China under Grant (61725205, 61876217, 3177167, 31671589, and 31371533), the Key Laboratory of Agricultural Electronic Commerce, Ministry of Agriculture of China under Grant (AEC2018003), the Anhui Foundation for Science and Technology Major Project under Grant (16030701092 and 18030901034), the 2016 Anhui Foundation for Natural Science Major Project of the Higher Education Institutions under grant (kJ2016A836), and the Hefei Major R&D Projects of Key Technologies under grant (J2018G14).

References

1. Z. Xia, X. Wang, X. Sun, Q. Wang, A secure and dynamic multi-keyword ranked search scheme over encrypted cloud data. IEEE Trans. Parallel Distrib. Syst. **27**(2), 340–352 (2015)
2. Z. Fu, X. Wu, Q. Wang, K. Ren, Enabling central keyword-based semantic extension search over encrypted outsourced data. IEEE Trans. Inf. Forensics Secur. **12**(12), 2986–2997 (2017)

3. M. Lease, On quality control and machine learning in crowdsourcing, in *Proceedings of the 11ᵗʰAAAI Conference on Human Computation* (AAAI, Menlo Park, 2011), pp. 97–102

4. J. Zhang, X. Wu, V.S. Sheng, Learning from crowdsourcing labeled data: A survey. J. Artif. Intell. Rev. **46**(4), 543–576 (2016)

5. V.S. Sheng, F. Provost, P.G. Ipeirotis, Get another label? Improving data quality and data mining using multiple, noisy labelers, in *Proceedings of the 14ᵗʰACM SIGKDD International Conference on Knowledge Discovery and Data Mining* (ACM, New York, 2008), pp. 614–622

6. G. Li, J. Wang, Y. Zheng, M.J. Franklin, Crowdsourced data management: a survey. IEEE Trans. Knowl. Data Eng. **28**(9), 2296–2319 (2016)

7. Mturk. https://www.mturk.com/. 2017

8. Crowdflower. http://www.crowdflower.com/. 2017

9. G. Li, Y. Zheng, J. Fan, J. Wang, R. Cheng, Crowdsourced data management: overview and challenges, in *Proceedings of the 2017 ACM International Conference on Management of Data* (ACM, New York, 2017), pp. 1711–1716

10. Q. Wang, V.S. Sheng, C. Hu, Keyphrase extraction using sequential pattern mining and entropy, in *Proceedings of the 2017 IEEE International Conference on Big Knowledge* (IEEE, Piscataway, 2017), pp. 88–95

11. I.H. Witten, G.W. Paynter, E. Frank, C. Gutwin, C.G. Nevill-Manning, KEA: practical automatic keyphrase extraction, in *Proceedings of the 4th ACM Conference on Digital libraries* (ACM, New York, 1999), pp. 1–23

12. G. Ercan, I. Cicekli, Using lexical chains for keyword extraction. Inf. Process. Manag. **43**(6), 1705–1714 (2007)

13. S. Xu, S. Yang, C.M. Lau, Keyword extraction and headline generation using novel word feature, in *Proceedings of the 24th AAAI Conference on Artificial Intelligence* (AAAI, Menlo Park, 2010), pp. 1461–1466

14. K.S. Hasan, V. Ng, Automatic keyphrase extraction: a survey of the state of the art. in *Proceedings of the 52nd Annual Meeting of the Association for Computational Linguistics*, pp. 1262–1273, 2014

15. Z. Fu, X. Wu, C. Guan, X. Sun, K. Ren, Toward efficient multi-keyword fuzzy search over encrypted outsourced data with accuracy improvement. IEEE Trans. Inf. Forensics Secur. **11**(12), 2706–2716 (2016)

16. Z. Fu, K. Ren, J. Shu, X. Sun, F. Huang, Enabling personalized search over encrypted outsourced data with efficiency improvement. IEEE Trans. Parallel Distrib. Syst. **27**(9), 2546–2559 (2016)

17. R. Agrawal, R. Srikant, Mining sequential patterns, in *Proceedings of the 11th International Conference on Data Engineering* (IEEE, Piscataway, 1995), pp. 3–14

18. G. Li, C. Chai, J. Fan, X. Weng, J. Li, Y. Zheng, CDB: optimizing queries with crowd-based selections and joins, in *Proceedings of the 2017 ACM International Conference on Management of Data* (ACM, New York, 2017), pp. 1463–1478

19. L. von Ahn, L. Dabbish, Labeling images with a computer game, in *Proceedings of the SIGCHI Conference on Human Factors in Computing Systems* (ACM, New York, 2004), pp. 319–326

20. Y. Zheng, G. Li, Y. Li, C. Shan, R. Cheng, Truth inference in crowdsourcing: Is the problem solved? Proc. Vldb Endowment **10**(5), 541–552 (2017)

21. Q. Wang, V.S. Sheng, Z. Liu, Exploring methods of assessing influence relevance of news articles. in *Proceedings of the 4th ICCCS(6)*, pp. 525–536, 2018

22. C. Shannon, A mathematical theory of communication. Bell Syst. Tech. J. **27**, 379–423 (1948)

23. INSPEC. https://github.com/snkim/AutomaticKeyphraseExtraction

Weakly Supervised Deep Learning for Objects Detection from Images

Jianfang Shi, Xiaohui Yuan, Mohamed Elhoseny, and Xiaojing Yuan

Abstract Training of a convolutional neural network for object detection requires a large number of images with pixel-level annotations. Weakly supervised learning uses image-level labels to circumvent the issue of lack of semantic examples, which remains an open challenging. This paper proposes a cascaded deep network architecture that leverages the class activation mapping with global average pooling. The first stage of this architecture learns to infer object localization maps based on the image-level annotations, which generates bounding boxes of objects in every image. These image patches are adhesion areas in the original image. In the second stage, the image patches are used to train the detection network. Experiments are conducted using the PASCAL VOC 2012 datasets. Our proposed method obtains a mean average precision of 87.2% and demonstrates a competitive performance of classification performance with respect to the state-of-the-art methods. In the evaluation of object localization, the recall of our method is improved by 9%.

Keywords Weakly supervised learning · Neural networks · Deep learning · Object detection

J. Shi
College of Information and Computer, Taiyuan University of Technology, Taiyuan, Shanxi, China

X. Yuan (✉)
Department of Computer Science and Engineering, University of North Texas, Denton, TX, USA
e-mail: xiaohui.yuan@unt.edu

M. Elhoseny
Faculty of Computers and Information, Mansoura University, Mansoura, Egypt

X. Yuan
University of Houston, Houston, TX, USA

1 Introduction

Pixel-level annotation of images for object detection is extremely time consuming and is the bottleneck of the deep learning methods such as convolutional neural networks (CNNs). When new classes are added, the previous annotations require revision. Considerable attention has been drawn to weakly supervised learning for semantic segmentation, in which much less number of training examples are used, e.g., Scribbles [1–3], bounding box [3–6], clicks [7], and image-level [3, 4].

Zhou et al. [8] highlighted that the convolutional filters in CNNs have specific receptive field and activation image pattern, and as the levels increase, the selectivity and semantics of neurons become more obvious. These object detectors spontaneously emerge within CNNs, and there is no object-level supervision in CNN training itself. Oquab et al. [9] employed global max-pooling (GMP) to solve weakly supervised prediction localization using CNN, which avoids the need for detailed object-level annotations. A common theme in these methods is that CNNs training relies on SIFT or HOG features. Lin et al. [10] employed network in network scheme and replaced global average pooling (GAP) with the full connection. It was demonstrated that reducing the numbers of fully connected layers results in the blocking of a large number of parameters without sacrificing classification accuracy. Zhou et al. [11] revisited the global average pooling and devised a method that explicitly enables the convolutional neural network to have a remarkable localization ability despite being trained on image-level labels. Papandreou et al. [4] extended DeepLab, in which the training data use the bounding box and image-level labels as markers are studied. Expectation–maximization algorithm was used to estimate the categories of unlabeled pixels and CNNs parameters. Simonyan et al. [12] used the GraphCut segmentation method on top of class saliency map, which is attempted to capture the continuity cue from color. Recent studies have shown that the GAP as a structural regularization strategy helps to avoid over-fitting to the training data and achieves a better localization performance [11]. Zhou et al. [11] proposed a class activation mapping (CAM) that visualizes the prediction scores and highlights the discriminative object part. The early research on weakly supervised object detection focuses mostly on computing CAM to obtain localization maps. The methods find parts of images with a visually consistent appearance to the training images that locate multiple objects of a category in the same region. The localization map generated with CAM usually results in unseparated multiple objects in the same category and weak supervision as a means of producing labels for each object has only been embraced to a limited degree. To address this issue, a new approach is proposed in this paper.

Our idea is to create a dual convolutional neural network architecture for object classification of complex scenarios that contain one or more foreground objects [13–17]. The method derives a better appearance model from images with multiple objects using the image-level training dataset. The networks are designed to jointly learn multiple, different but related tasks. The main contribution of this work is improving multiple object detection with weak supervision. The dual CNN architecture consists of two stages. In the first stage, the network learns to infer object localization, which produces bounding boxes of the candidates, i.e., the region of interests. The regions of interest are used to train the second network for classifying the class of the object in the image patch. By localizing the objects in the image followed by classification, the dual neural network assures coherence and labeling for the detection of multiple objects and gets a clear distinction between objects of the same category.

The rest of this paper is organized as follows: Section 2 presents our proposed method in detail. Section 3 discusses the experimental results and evaluates the performance with respect to state-of-the-art methods. Section 4 concludes this paper with a summary.

2 Cascaded Deep Network Architecture

Our proposed dual CNN architecture is trained to learn multiple class annotations using image-level labels. The labels of training data are represented with a string of 0s and 1s, where "1" indicates the existence of an object in that type and "0" indicates the absence of such type of objects in the image. A label vector is expressed as follows:

$$y = [y_1, y_2, \ldots, y_c], \quad y_c \in \{0, 1\}$$

The PASCAL VOC 2012 dataset [18] is used as the training data in this work, which consists of 20 categories. y_c indicates which category is included in the image. For example, a vector [0,1,0,0,0,0,0,0,0,0,0,1,0,0,0,0,1,0,0,0] indicates that this image includes objects of three classes (marked with ones). We converted multi-label in vector format to 1×20 binary images. The loss function is as follows:

$$E = -\frac{1}{N} \sum_{n=1}^{N} p_n \log \hat{p}_n + (1 - p_n) p_n \log \left(1 - \hat{p}_n\right) \tag{1}$$

The most commonly used data is the data type in Caffe [19], and the data type requires the input data to be lmdb or leveldb. Generating efficient data input involves data conversion. Original images and multi-label images were converted to a format of lmdb. Before original images and multi-label images were converted, they were subjected to the same random shuffle. Two LMDBs are established separately to process images and labels. Before images and multi-label are generated, they are subjected to the same random shuffle.

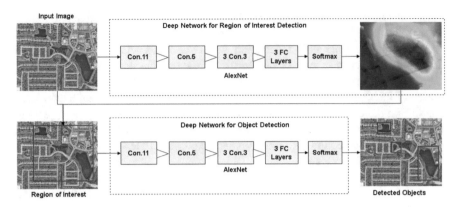

Fig. 1 The flowchart of the proposed dual CNN architecture for weakly supervised object detection

Our proposed dual neural network architecture for weakly supervised objects detection is as follows:

1. The first CNN learns to derive object localization maps using the image-level labels. This CNN generates a number of bounding boxes for objects in a given image and image patches are cropped and used for the processing in the next CNN.
2. The second CNN learns multiple object classes and performs classification of the image patches to achieve object detection in images.

Our method extends the approaches to compute CAMs of training images as presented in [1]. It is typically highlighted the local region of interests of objects. Furthermore, the region of interest is employed to generate sources of the next stage in this approach. As shown in Fig. 1, we trained a CNN for classification and obtained image localization by CAM. We compute a weighted sum of the feature maps of the last convolutional layer to obtain our class activation maps. Given the trained network, the CAM for a class c, denoted as M_c, is computed as follows:

$$M_c (x, y) = W_c^T f (x, y) \tag{2}$$

where W_c is the classification weights associated with a class c and $f(x, y)$ gives the feature value at a pixel (x, y) on the feature map of the last convolution layer (conv6).

The architecture of a classification network is based on AlexNet [20] with global average pooling (GAP), which is followed by a fully connected layer, and trained with image-level labels. The process receives interesting image regions by highlighting the weights of the output layer on the convolutional feature maps. For the method that relies solely on CAM to obtain object localization, separating multiple objects of the same category is challenging. Our method utilizes two CNNs following a cascaded structure.

The CAMs of the first network generate bounding boxes and a thresholding technique is employed to create the cropped images. With appropriately selected thresholds, bounding boxes of objects in an image are detected. The bounding box that yields the largest connected component in the segmentation map is taken. The bounding boxes were mapped on the original images. As shown in the localization image of stage 1 of Fig. 1, two objects of the same category (people) were in the same area. The result shows the output of multiple objects in a single image is very challenging. The coordinates of the bounding boxes were used to crop the area of interest from the original image to get the cropped images.

The confidence for classification of the new training images for a category is obtained in the previous model. The labels for the second stage are regenerated by that confidence. That is, the training images with confidence greater than 0.5 had a label of 1 for class c and 0 otherwise. In the second stage, we converted multi-label to a vector with the above method. Because vector is the form of the label when the network was trained, the multi-classification problem is broken down into multiple binary classification problems, finally, we used 0.5 as the critical confidence.

These cropped images are the adhesion areas in the original image. The second network uses the results (cropped images and new ground-truth labels) of the first stage as a supervision signal during training. The second deep network follows the same network structure as the one in the previous stage. The localization images were obtained by CAM. It could be clearly displayed from the localization image of stage 2 in Fig. 1 that the two people who were originally positioned in the same area are well separated. The result indicated our proposed method solves this practical problem partly.

3 Experimental Results

3.1 Dataset and Settings

The PASCAL VOC 2012 dataset is used in our experiments. The dataset consists of 20 categories. In our experiments, we use 5157 images for training and 5823 images for validation. The image-level labels of the objects are used.

The two CNNs in the framework are programmed with AlexNet. Three layers are added to the backbone network: a 3×3 convolution kernel with 1024 channels for a better adaptation to the target, a global average pooling layer for feature map aggregation, and a fully connected layer for classification. The networks use cross-entropy as the loss function. The networks are pre-trained on ImageNet [21] dataset, the pre-trained model obtained from the webpage named Caffe model Zoo, the URL of the webpage is http://caffe.berkeleyvision.org/model_zoo.html.

The computer operates on Windows 7 with Intel Core i7-2600, Tesla k20m GPU, and 4 GB memory. The initial learning rate for the CNNs is 0.01, and the network is trained with 60 epochs. The multi-label CNN architecture is implemented using Caffe framework, the batch size is 64.

3.2   *Performance*

In single-label image classification tasks, the accuracy of classification was generally measured using top-1 accuracy or top-5 accuracy. The evaluation method of the multi-label image classification task generally adopts mean average precision (mAP).

Average precision (AP) is computed as follows:

$$AP = \int_0^1 PRdr \tag{3}$$

where P denotes precision and R denotes recall.

The confidence scores of all test images were obtained by the model. The confidence scores of each class (such as a car) were saved to a file and sorted in descending order. In this order, the number N of selected pictures was gradually increased. Then calculate the recall and precision under each N. In this process, a confidence score greater than 0.5 was considered to be a true positive sample and we can get the max precision. Finally, per-class AP was obtained (Fig. 2).

If the model was required to be good or bad in all classes, the average value of the AP corresponding to all classes was calculated, which is called mAP.

$$mAP = \frac{1}{|QR|} \sum_{q \in QR} AP(q) \tag{4}$$

As shown in Table 1, the results for object classification on PASCAL VOC 2012 val sets are average precision and mAP measurement. It is clear that using the weakly supervision labels a better performance among CAM is achieved. The accuracy of the classification of our method (87.2%) is competitive in comparison with CAM (84.4%).

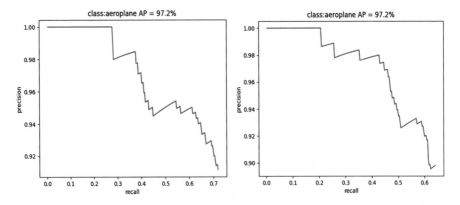

Fig. 2 The recall–precision curves for AlexNet for the aeroplane class

Table 1 Pre-class AP and mAP (%) on the PASCAL VOC 2012 validation set

Class	AlexNet	CAM	Our method
Aeroplane	97.2	97.2	**98.1**
Bicycle	97	96.9	**99.2**
Bird	**95.3**	88.1	93.5
Boat	**95.7**	88.5	94.2
Bottle	51.2	61	**65**
Bus	**99.4**	98.9	99.1
Car	**87.9**	86.4	87.3
Cat	92.7	90	**93.1**
Chair	**77.7**	69.2	76
Cow	**81.2**	67.6	78.6
Dining table	75.5	71.7	**78.7**
Dog	85.7	85.4	**86.2**
Horse	**90.6**	82.7	83.6
Motorbike	**99**	98.6	97.4
Person	95.1	94.9	**95.3**
Potted plant	**79.1**	76.7	77.8
Sheep	**93.2**	89.3	90.3
Sofa	**66.3**	59	62.4
Train	97.6	97.5	**97.8**
TV monitor	89	88.9	**90.2**
mAP	**87.3**	84.4	87.2

The numbers in boldface highlight the best performance for the classes

3.3 Performance of Detection

Intersection over Union (IoU) is a metric that measures the accuracy of detecting objects in a specific dataset. To measure objects of any size and shape we need the ground-truth bounding boxes and the prediction bounding boxes obtained by the algorithm (the bounding boxes were generated with less error by setting an appropriate threshold). IoU is calculated as follows:

$$IoU = \frac{\#A_o}{\#A_u} \tag{5}$$

where $\#(A)$ denotes the number of pixels in a region denoted with A, A_o denotes the area of overlap, and A_u denotes the area of union. A graphical illustration is shown in Fig. 3.

As shown in Table 2, the mIoU of the method outperforms the CAM by 2.1%. For the classes of "bird," "bottle," "cow," and "TV," there exist multiple objects of the same category. Our method outperforms CAM by 10%. The results demonstrate the proposed method addresses the multi-object problem and gets an improved distinction between objects of the same category in an image.

Fig. 3 Graphical representation of the formula. (**a**) Area of overlap. (**b**) Area of union

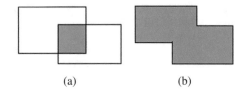

(a)   (b)

Table 2 mIoU of our method and CAM

Class	Aeroplane	Bicycle	Bird	Boat	Bottle	Bus	Car
CAM	48	46.2	34.2	**27.7**	15.6	**56.1**	36.6
Our method	**49.1**	**49.2**	**43.2**	27.6	**27.7**	55.6	**36.7**
Class	Cat	Chair	Cow	Table	Dog	Horse	Bike
CAM	**55.3**	21.2	47.8	31.9	**44.1**	**52.9**	**56.4**
Our method	54.1	**21.4**	**58.1**	**37.2**	39.1	51.5	55.1
Class	Person	Plant	Sheep	Sofa	Train	TV	**mIoU**
CAM	**35.8**	36.9	38	37.5	**46.8**	27.2	39.8
Our method	33.7	**38.6**	**38.9**	**42.6**	44.7	**33.3**	**41.9**

The results are reported in percentage
The numbers in boldface highlight the best performance for the classes

The recall and precision of localization instead of mIoU are used as the evaluation index in this work. If any of the bounding boxes in the image has confidence greater than 0.5 for a class, this image is a case for that class. In our comparison, stage 11 is the model used in the first stage. Specifically, the difference between this method and our method was this method still using the weights of the first stage to make predictions in the second stage.

During the experiment, we tried a variety of evaluation methods, such as the mIoU. The combination of recall and precision could better reflect the pros and cons of the required target. Because the target of this paper was to solve the problem of similar multi-objects adhesion in a single image, it was possible that a certain class cannot mark the complete object position by using CAM, the localization was more concentrated on the most interesting area. If mIoU or another method was used as the only evaluation index, the obtained result would have a large error. Precision and recall are mutually influential. Ideally, both precision and recall are high, but in general, recall increases, while precision decreases and vice versa. In this work, if the adjacent targets are separated, the value of true positive could be significantly improved. That is, the recall is improved.

In Table 3, compared to CAM, the recall of stage 11 increased by 4.4% and the recall of our method increased by 8.5%. For our method, approximately 4.1% improvement can be obtained over stage 11. It can be observed that our deep neural network architecture achieved improved appearance model and feature representation for multiple classes. In addition, the proposed approach is trained in an end-to-end learning fashion of deep neural networks.

In Fig. 4, we overlay the localization maps generated by the visualization with the original image to illustrate the advantages of the method for dealing with multi-

Table 3 Detection performance comparison on the PASCAL VOC 2012 val set with recall and precision metrics

Class	CAM		Stage 11		Our method	
	Recall	Precision	Recall	Precision	Recall	Precision
Aeroplane	36	64.4	41.4	56.1	**45.9**	**67.8**
Bicycle	21.6	35.3	23.5	**62.9**	26.5	60.1
Bird	13.5	30.8	**28.5**	28.9	27.6	**33.7**
Boat	6.1	20	10.4	**25.2**	10.9	21.1
Bottle	1.5	7.1	0.6	9.2	**12.5**	**9.6**
Bus	39.4	**67**	40.6	64.6	**46.9**	64.3
Car	10.8	28.5	15	**31**	15.3	27.6
Cat	40.9	**48.6**	**64**	47.4	62.7	48.5
Chair	2.5	**8**	2.7	4.9	**5.2**	7.5
Cow	8.1	**39.4**	7.8	36.7	**9.6**	33.3
Dining table	7	21.5	7	**23.4**	18.2	17.5
Dog	26.8	**47.6**	34.1	46.2	35.5	27.3
Horse	13.7	**58**	22.6	57.1	26.9	50
Motorbike	26.9	**73.2**	27.6	58.2	**37.8**	61.4
Person	19.6	35.5	17.5	**40.1**	25.2	28.7
Potted plant	3.5	**27.1**	3.8	23.4	**5.5**	21.4
Sheep	5.4	32.5	10	41.9	**16.3**	**48.7**
Sofa	5.9	17.4	4.4	15.5	**11.3**	**19.4**
Train	33.7	**60.3**	**50.6**	51.4	48.3	55
TV monitor	12.3	**25**	11.5	20.4	**17.7**	19.9
Average	16.8	**37.4**	21.2	37.2	**25.3**	36

The numbers in boldface highlight the best performance for the classes

objects problems. In Fig. 4, the areas in red depict the potential and detected object areas. The left column, Fig. 4a, shows the localization maps obtained from the CAM. As depicted in this figure, the multi-objects position is unaddressed. The middle column of Fig. 4 (stage 11) depicts the localization maps, i.e., the region of interest and the multi-object detection results start to improve. The localization maps shown in the right column, Fig. 4c, that are obtained by our proposed method yielded the best object detection results. These results demonstrated that the proposed approach achieved improved multi-object detection performance.

3.4 Training Efficiency and Network Convergence

Figure 5 illustrates the loss of constructing CNNs over the training iterations. In our experiments, the network output a loss every 100 iterations during the training phase. Figure 5a depicts the loss of training the first deep network to detect the bounding box of objects in an image. Figure 5b depicts the loss of training

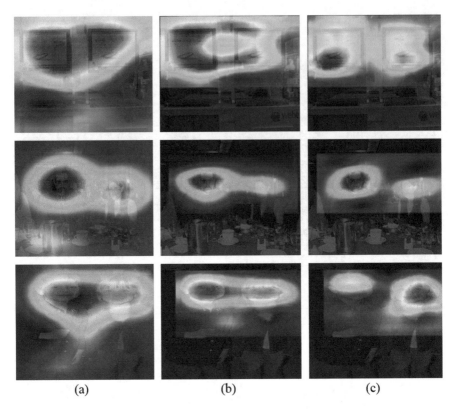

(a)   (b)   (c)

Fig. 4 The resulted localization maps superimposed to the input images. (**a**) CAM, (**b**) stage 11, (**c**) our method

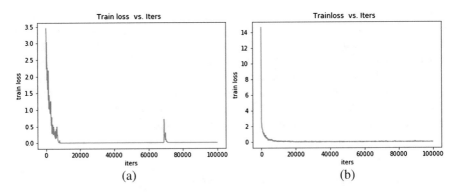

Fig. 5 The training loss of the two CNNs. (**a**) Loss of the CNNs to generate bounding box. (**b**) Loss of the CNN for object recognition

the second deep network to classify each region of interest extracted from the input image. The results show that both networks converged after around 10,000 iterations. Note that this is an average of ten repetitions and the loss curves represent a general trend. In comparison to the first CNNs, the network for image patch classification converged faster. It is, hence, acceptable to stop training much earlier if efficiency is a concern.

4 Conclusion

In this paper, we proposed a weakly supervised learning method based on a dual CNN architecture for multiple objects detection from images. Our method employs CAM for CNNs with GAP, which created two models for detecting a region of interest and object classification. We evaluated our method using the PASCAL VOC 2012 validation dataset.

Our experimental results demonstrate that the accuracy of the classification of our proposed method is competitive in comparison with AlexNet. The mIoU of the method outperforms the CAM by 2.1%. The advantage is much more significant in dealing with cases of multiple objects of the same category, in which our method outperforms CAM by 10%. The results demonstrate the proposed method addresses the multi-object problem and gets an improved distinction between objects of the same category in an image. It is also demonstrated that the network training was efficient and a convergence was reached around 10,000 iterations.

References

1. B. Zhou, A. Khosla, A. Lapedriza, A. Oliva, A. Torralba, Learning deep features for discriminative localization. in *Computer Vision and Pattern Recognition*, 2016
2. G. Papandreou, L.C. Chen, K. Murphy, A.L. Yuille, Weakly-and semi-supervised learning of a CNN for semantic image segmentation. in *International Conference on Computer Vision*, 2015
3. A. Chaudhry, P.K. Dokania, P.H.S. Torr, Discovering class-specific pixels for weakly supervised semantic segmentation. in *Computer Vision and Pattern Recognition*, 2017
4. N. Liu, J. Han, DHSNet: deep hierarchical saliency network for salient object detection. in *IEEE Conference on Computer Vision and Pattern Recognition*, pp. 678–686, 2016
5. A. Kolesnikov, C.H. Lampert, Seed expand and constrain: three principles for weakly-supervised. In *European Conference on Computer Vision*, pp. 695–711, 2016
6. H. Zhang, Z. Kyaw, J. Yu, S.F. Chang, PPR-FCN: weakly supervised visual relation detection via parallel pairwise R-FCN. in *IEEE International Conference on Computer Vision*, pp. 4243–4251, 2017
7. K. He, X. Zhang, S. Ren, J. Sun, Spatial pyramid pooling in deep convolutional networks for visual recognition. In *European Conference on Computer Vision*, pp. 346–361, 2014
8. M. Tang, A. Djelouah, F. Perazzi, Y. Boykov, C. Schroers, Normalized cut loss for weakly supervised CNN segment. in *Computer Vision and Pattern Recognition*, 2018

9. J. Long, E. Shelhamer, T. Darrell. Fully convolutional networks for semantic segmentation. in *Proceedings of the IEEE Conference on Computer Vision and Pattern Recognition*, pp. 3431–3440, 2015

10. J. Dai, K. He, J. Sun, BoxSup: exploiting bounding boxes to supervise convolutional networks for semantic segmentation. in *IEEE International Conference on Computer Vision*, pp. 1635–1643, 2015

11. P.O. Pinheiro, R. Collobert. Weakly supervised semantic segmentation with convolutional networks. in *Computer Vision and Pattern Recognition*, 2015

12. D. Lin, J. Dai, J. Jia, K. He, J. Sun, ScribbleSup: scribble-supervised convolutional networks for semantic segmentation. in *Proceedings of the IEEE Conference on Computer Vision and Pattern Recognition*, pp. 3159–3167, 2016

13. N. Krishnaraj, M. Elhoseny, M. Thenmozhi, M.M. Selim, K. Shankar, Deep learning model for real-time image compression in Internet of Underwater Things (IoUT). J. Real-Time Image Process. 2019. https://doi.org/10.1007/s11554-019-00879-6

14. M. Elhoseny, G.-B. Bian, S.K. Lakshmanaprabu, K. Shankar, A.K. Singh, W. Wu, Effective features to classify ovarian cancer data in internet of medical things. Comput. Netw. **159**, 147–156 (2019)

15. B.S. Murugan, M. Elhoseny, K. Shankar, J. Uthayakumar, Region-based scalable smart system for anomaly detection in pedestrian walkways. Comput. Electr. Eng. **75**, 146–160 (2019)

16. K. Shankar, M. Elhoseny, S.K. Lakshmanaprabu, M. Ilayaraja, R.M. Vidhyavathi, M. Alkhambashi, Optimal feature level fusion based ANFIS classifier for brain MRI image classification. Concurrency Comput. Pract. Exp. 2018. https://doi.org/10.1002/cpe.4887

17. X. Yuan, D. Li, D. Mohapatra, M. Elhoseny, Automatic removal of complex shadows from indoor videos using transfer learning and dynamic thresholding. Comput. Electr. Eng. **70**, 813–825 (2018)

18. P. Sermanet, D. Eigen, X. Zhang, M. Mathieu, R. Fergus, Y. LeCun, OverFeat: integrated recognition, localization and detection using convolutional networks. in *IEEE Conference on Computer Vision and Pattern Recognition*, 2013

19. T. Durand, T. Mordan, N. Thome, M. Cord, WILDCAT: weakly supervised learning of deep ConvNets for image classification, pointwise localization, and segmentation. in *IEEE Conference on Computer Vision and Pattern Recognition*, pp. 5957–5966, 2017

20. Y. Wei, X. Liang, Y. Chen, X. Shen, M.-M. Cheng, J. Feng, Y. Zhao, S. Yan, STC: a simple to complex framework for weakly supervised semantic segmentation. in *IEEE Transactions on Pattern Analysis and Machine Intelligence*, 2016

21. Y. Wei, J. Feng, X. Liang, M.M. Cheng, Y. Zhao, S. Yan, Object region mining with adversarial erasing: a simple classification to semantic segmentation approach. in *Conference on Computer Vision and Pattern Recognition*, 2017

Index

© Springer Nature Switzerland AG 2020
X. Yuan, M. Elhoseny (eds.), *Urban Intelligence and Applications*, Studies in
Distributed Intelligence, https://doi.org/10.1007/978-3-030-45099-1

Printed in the United States
by Baker & Taylor Publisher Services